EXXON AND THE CRANDON MINE CONTROVERSY

Michael O'Brien

Badger Books LLC, Middleton, Wisconsin, USA

Badger Books LLC
1600 N. High Point Rd.
Middleton, Wisconsin 53562, USA
Tel: 1-800-928-2372
Email: books@badgerbooks.com
Website: www.badgerbooks.com

ACKNOWLEDGMENTS

Thanks to Julie Brunette for guiding me through the papers of the Wisconsin Department of Natural Resources; to Arden Rice and Rose Arnold for doing the same with the papers of the Wisconsin Legislative Reference Bureau.

A special thanks to Tina Van Zile of the Sokaogon Chippewa Community and to Glenn Reynolds for their assistance, particularly for arranging my study of the Nicolet Minerals Company papers.

Many persons involved in the mine controversy loaned or gave me their personal papers. For their cooperation I am indebted to Linda Bochert, J. Wiley Bragg, Al Gedicks, Zolton Grossman, Barry Hansen, Toni Harris, Sandy Lyon, Mary Lou Munts, Peter Peshek, George Rock, Sonny and Maureen Wreczycki, and Carl Zichella.

Thanks for the general assistance of Larry Nesper, Mike Monte, John Mutter, and Shelley Spaude.

I owe special thanks to staff members at the University of Wisconsin-Fox Valley who provided assistance over several years. They are Becky Hoffman, Sarah White, Patricia Warmbrunn, Christine Chamness, and Max Schultz. Almost daily for three years, Kathy Skubal expertly processed mountains of written drafts, always doing the work with patience and humor. Dean Jim Perry endorsed my project from the beginning. Colleagues Marc Sackman, Joy Perry, Dave Hager, Richard Krupnow, Kenn Forsythe, and Dubear Kroening reviewed the manuscript and offered suggestions.

My wife, Sally, graciously endured her husband's seventh book with patience, encouragement, and love. Thanks to my children, Tim, Sean, Jeremy, and Carey for all the love and enjoyment they bring me.

O'Brien Biography

MICHAEL O'BRIEN, a native of Green Bay, Wisconsin, received his B.A. in 1965 from the University of Notre Dame. He did his graduate work at the University of Wisconsin-Madison, receiving his Ph.D. in 1971. He taught for thirty-two years at the University of Wisconsin-Fox Valley, Menasha, Wisconsin, until his recent retirement.

O'Brien is the author of McCarthy and McCarthyism in Wisconsin (University of Missouri Press, 1981), Vince: A Personal Biography of Vince Lombardi (William Morrow, 1987), Senator Philip Hart: the Conscience of the Senate (Michigan State University Press, 1996), Hesburgh: A Biography of Fr. Theodore Hesburgh (The Catholic University Press, 1998), and Not an Ordinary Joe: A Biography of Joe Paterno (Rutledge Hill Press, 1998). His sixth book, John F. Kennedy: A Biography, was published by St. Martin's Press in March 2005.

O'Brien's writing has earned top awards from Choice Magazine, the Wisconsin Library Association, the National Catholic Press Association, and the Wisconsin Magazine of History.

Recently he published A Writing and Teaching Life: A Memoir (Publish America, 2007).

Table of Contents

INTRODUCTION

In 1976 news about a huge mineral discovery and proposed mine just south of Crandon, Wisconsin, worried the Menominee and the Potawatomi Indian tribes whose reservations were nearby, but the mine directly endangered the Sokaogon Chippewa tribe because it bordered their reservation at Mole Lake.

The Chippewa, also known as the Ojibwa, traced their origin to Algonquian-speaking Indians who lived near the Atlantic coast not far from the mouth of the St. Lawrence River. Gradually moving west, a small band of Sokaogon Chippewa settled at Mole Lake, Wisconsin. In 1806 the Sioux encroached on Mole Lake's rice beds but were repelled in a furious bloody battle that killed five hundred warriors and left the area a sacred burial ground, now threatened by the mine project. Sokaogon oral tradition, passed down through generations, maintained that in the 1850s the federal government promised them a twelve square mile reservation near Mole Lake, but because the document confirming the arrangement was lost in a shipwreck, the Sokaogon had no proof of ownership. "For over 80 years, the Sokaogons continued to occupy their traditional territory [at Mole Lake], but their identity as a separate band was ignored by the United States," said one authority.[1]

In 1917 Sokaogon Chief Edward Ackley, returning home from military service in World War I, described the plight of his people:

I came home...to find my people...in a state of bewilderment, like sheep without a leader. They had nowhere to live, some were sick from hunger, and some were freezing from lack of wood for their wigwams and tepees. They were a destitute band of Indians with no land to live. As the white settlers bought the land here in this area, they would tell the Indians to move; and that they had done for many years.

Two years later, W.S. Coleman, an inspector for the U.S. Department of Interior, found the conditions at Mole Lake deeply disturbing. One hundred-and-fifty Sokaogon were huddled on one acre of land, "living in tar-paper shacks and eating the handouts of

1

local whites. They had neither livestock nor crops, nor even enough clothes to protect them during the sub-zero winters."

The Indian Reorganization Act of 1934 finally recognized the Mole Lake band as a tribe and by 1939 the Mole Lake Reservation was complete and their land base was established. In 1976 after purchasing additional land, the Mole Lake Reservation consisted of 1700 contiguous acres located in the Town of Nashville in Forest County, Wisconsin. [2]

Still a poor tribe, its government occupied an office the size of a two-car garage and conducted business on a yearly operating budget of $900. According to the tribal secretary, Daniel Polar, tribal members lived in "poor housing," received "poor health care," and had an unemployment rate of about 34%.

"Alcohol abuse and alcoholism are probably the most serious health problems among the Sokaogon," one study claimed. "Problems connected with alcohol are thought to affect 50% or more of the adult population and many young people." Health officials reported depression to be the most common form of mental illness among the Sokaogon.

Several factors hampering economic development on the reservation included the lack of entrepreneurial skill and the tribe's reluctance to exploit natural resources. "Employment and training opportunities in the past have not facilitated the tribe's participation in the dominant economy without migration to urban areas," said another study.[3]

Tribal members lacked the work experience, the education and the skills needed for employment at a mining construction site. "The jobs [tribal members] might get [at the mine] are sweeping the floor or driving a truck," said Mole Lake's Wayne LaBine. "The high paying jobs just aren't there for [us]," he said, predicting that skilled union mine workers from out of state would end up with the best positions.[4]

Water was fundamental to sustaining the tribe's physical and economic well-being and was valued in the cultural and spiritual beliefs of the tribe. Among several lakes within the reservation, the tribe prized Rice Lake, a shallow body of water covering 320 acres of

2

carefully guarded and cultivated rice beds called manomin (pronounced "men omen").

In the early 1500s the Chippewa's search for wild rice stimulated their migration from the Atlantic seaboard to the Great Lakes. "According to the migration story handed down through the years, the [Chippewa] people were told through a prophet that their creator, Gichi-Manidoo, wanted them to return to the Great Lakes where they had once lived. They were told to follow a sacred shell to seven stopping places and, eventually, to 'the food that grows on water.'" Their last stop was the rice beds at Mole Lake.

The rice sometimes provided thirty percent of the tribe's annual income. Money from selling surplus rice bought school clothes for the children, but first, the tribe always provided for the elderly and for others unable to gather their own food.[5]

Tribal Judge Fred Ackley described the process of gathering the rice. "One person slowly pushes a boat through shallow water while a second person gently taps the rice seeds loose into the boat. Then the rice is dried over an open fire, danced upon to shake off the outer shells, thrown into the air to dust off remaining particles, and then cleaned by hand." Rice meant "food" to Ackley. "It is my traditional way of life. It's like Americans and apple pie. It feeds my spirit."

Journalist Ron Seely of the <u>Wisconsin State Journal</u> reported on Ackley's tender relationship with the wild rice. Ackley remembered the sky in the fall "filled with ducks and geese coming in to rest on Rice Lake and feed in the rice beds." During harvesting, fires burned "so the big metal washtubs could be filled with the rice and placed on the fires to cure." "And you could smell it when it was roasting," he recalled. Eating the rice was delightful. "With just a little bit of a pound, you could feed a lot of people," Ackley said. "Put in a little bit of deer meat or duck....We didn't have cereal and instead my mother at night would put a handful of rice in a pan of water and in the morning it would be all puffed up. We'd put it in a bowl and put canned milk and maple syrup on it and eat it for breakfast."[6]

Dr. Nancy Lurie, Chief Curator of Anthropology at the Milwaukee Public Museum, pointed out the rice was much more than food.

It represents a culmination of the natural cycle in which the air, soil, water, sun, organic matter, and animal matter come together to provide the Indians with food. It is antithetical to their culture and traditional religious views to disturb or modify those factors solely for the purpose of providing one kind of food exclusively for human beings. The wild rice [is] also part of a total ecosystem which also sustained fish, wild birds, and wild animals.

"If the spiritual health of the water is neglected, or not maintained through respectful behavior and ceremony, it may lose its life-giving spirit, and the health and survival of plants, animals, and people will be in danger," noted Larry Nesper who studied the beliefs of the Sokaogon tribe.[7]

The Sokaogon Chippewa feared any mine operation; it could pollute the water and ruin the wild rice. Mining would drop the water level of Rice Lake and threaten the rice beds. It threatened a cherished river on the reservation as well. Less than a mile north of the mine site was Swamp Creek, clean enough "to support ecologically fussy trout." Swamp Creek flowed into Rice Lake, and then into the Wolf River.

"We can't move from here," observed LaBine. "[Non Indians] around us can move. We can't do that....This is our land, our water, our life. You can't put a price on life. We have to leave this place as good or better than we have it now for the sake of our children."[8]

CHAPTER ONE
REACHING CONSENSUS

In 1969 Exxon Corporation began a sulfide minerals exploration program in northern Wisconsin using airborne electromagnetic surveys. In June 1975 the method pinpointed an anomaly south of Crandon, and subsequent ground drilling tests located a huge mineral deposit. Exxon publicly announced the discovery on May 12, 1976.

The magnitude of Exxon's mine discovery near Crandon surprised many residents of northern Wisconsin, the state legislature, and the governor. Some folks thought the mine could endanger the Wolf River, the land and groundwater, and pondered its impact on the culture and religious beliefs of three nearby Indian tribes.

Some residents grew suspicious and pessimistic about having a giant corporation prowling in Wisconsin woodlands. Others expected the mine would create sorely needed high-paying jobs and greatly increase the tax base.

The ore body was about 5,000 feet long, 1,500 feet deep and 100 feet wide. When divided into two sections, one zone was predominantly 8.4% zinc, the other 0.7% copper, totaling an estimated 67.4 million tons. Two hundred feet of glacially deposited soil and rock covered the mineral resource. The quality of the mineral deposit was average. "The size made up for the lack of the high grade material," said one company official.[1]

Copper is one of the best conductors of electricity and is used in tubing, computers, and telephones. Zinc is primarily used as a coating on iron and steel to protect against corrosion. In the average automobile, seventeen pounds of zinc protect it from rust. The United States penny is 98% zinc with a copper coating.

The magnitude of the deposit shocked Wisconsin residents. A previous discovery by Kennecott Copper Company was only 1.5 million tons. "Until then, we had been plodding along, thinking new discoveries would be the size of Kennecott's," noted Robert H. Milbourne, an administrator in the Wisconsin Department of Revenue. "Exxon's announcement changed the whole ball game."[2]

For the northern Wisconsin region, the discovery was considered either a boon or a nightmare. The last industrial endeavor in the region had been a disaster when in the last quarter of the nineteenth century; approximately sixty billion board feet were derived from timber cut by lumbermen. The cutover region in northern Wisconsin was devastated. "Brush, limbs, tops, and unwanted logs littered the ground, which still contained millions of stumps," noted historian Robert Gough. "With less vegetation, runoff increased, lowering stream levels, drying up marshes, and creating flood problems." Agriculture in the region failed. A visitor to the area in 1921 noted that "One who has not seen 'cutover country' can hardly imagine the task of clearing its stumps. It's almost like asking a man to transform the war scarred fields around Verdun into fruitful farms."

In the 1950s, northern Wisconsin lost population. Poor employment opportunities led to the outward migration of young people. "Levels of education, job skills, and income in the north country were consistently below that of the state as a whole," said historian William Thompson in his history of the state. Northern Wisconsin needed jobs, but skeptics feared mining might foster the economic boom-or-bust days of the timber era. By the 1970s the region's economy was still anemic, relying on two industries – forest products and tourism.

The Crandon discovery was located in northern Wisconsin's Forest County (population 8,000 in 1976), about thirty miles east of Rhinelander and twenty miles south of the Michigan border. One of the state's poorest counties, it had the lowest per capita income in the state, an unemployment rate of 13.9%, and over a thousand people on welfare. Seventy percent of the county was under state or federal ownership and therefore not taxable.

The county didn't have a stoplight or a single fast-food restaurant. "Civilization in Forest County passes by in the steady cough of a fisherman's boat motor, in a bareling logger's truck, [and] winter's whir of snowmobiles." Tree-lined and beautiful, the county is adorned with 200 lakes and 480 miles of trout streams. During the summer tourist season, the county nearly doubles its population. In Forest County in 1976, 49.6% of homes were vacation homes, the fourth highest percentage in the state.[3]

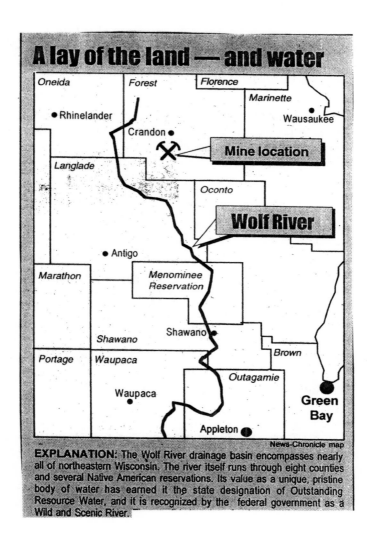

A lay of the land — and water

Oneida

Forest

Florence

Marinette

● Rhinelander

Wausaukee

Crandon ●

Mine location

Langlade

Oconto

Wolf River

● Antigo

Marathon

Menominee
Reservation

Shawano

Shawano ●

Shawano

Brown

Portage

Waupaca

Outagamie

Waupaca
●

**Green
Bay**

Appleton ●

News-Chronicle map

EXPLANATION: The Wolf River drainage basin encompasses nearly all of northeastern Wisconsin. The river itself runs through eight counties and several Native American reservations. Its value as a unique, pristine body of water has earned it the state designation of Outstanding Resource Water, and it is recognized by the federal government as a Wild and Scenic River.

Most of the residents of Crandon (population 1,582), the seat of the Forest County government, worked at the Connor Lumber Company in Laona. The only major industry in Crandon was the Bemis Manufacturing Company, maker of bowling pins and bed posts. The

7

promise of hundreds of mining jobs paying $10 or $12 an hour appealed to many local residents.

Yet there was fear of the mine's impact on nearby forests, lakes, and on the majestic Wolf River. No ordinary river, the Wolf was pristine, and on some days one could drink from it. "I'd rather drink this water, I'll tell ya," said nearby resident Herb Buettner, "than drink that stuff that comes out of the tap in Milwaukee."[4]

"Its name lingers in the heart and soul of every outdoorsman," said one historian about the Wolf River. The river originates from a series of springs, naturally exposed aquifers, marshes, wetlands, and depressions in Forest County. The federal government had designated twenty-four miles of the river as one of eight National Wild and Scenic Rivers, and the Wisconsin Department of Natural Resources had protected the Upper Wolf by purchasing nearly 80% of the river frontage at a cost of $10 million. Known for its high quality brown, brook, and rainbow trout and as vital habitat for sturgeon, the river also provides essential habitat for several endangered and threatened species including the bald eagle and osprey. Wild and angry with boulder-strewn stretches, the Wolf River topples down nearly seven hundred feet in a seventy-two mile span to Keshena Falls on the Menominee Indian Reservation. Walleye swim a hundred miles out of Lake Winnebago to spawn in it. "Game fish are hard put to withstand deteriorating conditions, including pollution, silt, destruction of spawning grounds."[5]

Would mining really endanger the Wolf and the environment near the river? "The extraction method," noted <u>Business Week</u>, "threatened to leave large amounts of waste rock that would have to be concealed." The waste rock, containing pyrites, is not dangerous when underground and dry, but when brought to the surface and exposed to moisture, the sulfides form sulfuric acid, creating a serious environmental threat. The acid leaches from tailings can contaminate surface and groundwater.

Exxon Corporation is one of the world's oldest companies, a leader in the oil industry since it was organized by John D. Rockefeller in 1882. By 1976 it was one of the largest corporations in the world. "Today, Exxon is well on the way to becoming a broadly based energy company, with substantial commitments in coal, synthetic fuels, and other ventures outside its traditional oil and

8

gas business," the company reported. It conducted business in nearly one hundred countries. "We explore for petroleum on six continents....Each day, about six million motorists stop at [our] 65,000 service stations."[6]

Exxon's initial announcement about the mine stirred public concern, and some state government officials were amazed at the company's lack of tact and public relations efforts. "I was in the Department of Natural Resources (DNR) building at the time the announcement came out," said Exxon geologist Edwarde May. "I've never seen so many bureaucrats bouncing off the walls in my life....That's not what a bureaucrat wants to hear. It was like a bolt of lightning." It was a jolt for Wisconsin's governor as well. Democratic Governor Patrick Lucey had not received the customary advance notice. "Lucey was dismayed," recalled J. Wiley Bragg, Exxon's public relations expert. "It was the wrong foot to start with in the state of Wisconsin." Governor Lucey worried about the effects of mining on the state and observed that "The majority of the people employed up there will still be employed in the tourist industry when the mine closes down, and we must be careful not to let the mining operations develop in a way that would close down the tourist industry." If Exxon Corporation's projections were correct, the Crandon zinc-copper discovery could be the largest economic boon to northern Wisconsin since the early lumbering operations, but, Lucey added, the pine forests of northern Wisconsin were wiped out in just thirty years by the lumber industry, "leaving the area barren and with no economic base....I would hope northern Wisconsin wouldn't get ripped off a second time."

"I'm confused just trying to tell you about it," Crandon's Mayor Gwinn Johnson said. "We don't know what's going to happen or, when it does happen, what problems will be brought up. It's like being out in the middle of the woods without a compass, and no stars in the sky."

In 1977, the daughter of an Exxon official living in Rhinelander returned home from high school crying. "They hate us, dad," she said. Her teacher had announced that northwoods people must "unite and send these (Exxon) people back home."[7]

Many residents were suspicious of Exxon, reported the New York Times. "Some believe the company is withholding information

9

because of fear of competitors. In the minds of some people, Exxon officials are city slickers trying to take advantage of country folk." Some felt "overwhelmed by the prospect of a multinational corporation determining their future." "Exxon will get what it wants," said one pessimistic resident. "Who are we to fight such a big company? We only hope we can get help from the state and federal governments to make certain Exxon doesn't rape the land, destroy our way of life, and leave."

Competition from other mining companies prompted Exxon to proceed quietly. News of successful discoveries usually attracted other exploration teams. "Most of Exxon's maneuvers have been cloaked in secrecy," noted Madison's Capital Times.[8]

Indeed, there was much secrecy in the way Exxon bought up property near the mine site, cloaking their purchases with confidentiality agreements. In what Crandon's city attorney described as a "questionable deal," in November, 1975, three supervisors for the Town of Nashville secretly leased to Exxon the mineral rights to land beneath some town roads.

When Exxon proposed to lease forty acres of public land, it asked the Wisconsin Board of Public Land Commissioners to keep the proposal confidential. Wisconsin's Secretary of State Douglas LaFollette protested. "This time around when the mining industry comes to northern Wisconsin, we have to make sure they don't just rip up the land and rip off the people and leave nothing behind when they abandon our state," LaFollette said. People remembered the disastrous results of the last big boom in northern Wisconsin, he continued. "When the big corporations pulled out, they had taken the wealth of the land and millions of dollars in profits, and they left behind unemployed workers, scarred land and economic depression. We should never forget what these corporations are, and it is wishful thinking to assume things will be different this time around."[9]

Actually Wisconsin has a mining history. The state's official seal displays a miner, mining equipment, and pig lead. Some early miners in the state lived in abandoned mines, giving Wisconsin a subtitle, the Badger State. However early laws regulating metallic mining were ineffective. The state needed new updated laws. The task was complex, and legislative leaders settled on a novel approach to establish the laws. They delegated to a small group including Exxon

10

the responsibility of hammering out detailed legislation. This formulating process was not a legislative process, but its results easily passed the legislature. While initially praised by some public officials, critics subsequently questioned the fairness of the process and the mining laws it produced. Sessions, dominated by Exxon, were closed. Indian tribes and citizens' groups who opposed the new rules had been ignored.

In the early 1960s Wisconsin's government officials began studying the environmental effects of mining and focused on the impacts of mining sand, gravel, and stone. But were these laws effective when addressing environmental impacts caused by metallic minerals, such as the ore deposits discovered by Kennecott and Exxon?

In 1976, a nine-member bi-partisan Citizen Advisory Committee for the Public Intervener held its first retreat in Bayfield County at the cottage of one of its members. Among the guests invited to address the group was Roscoe Churchill of Ladysmith who argued that Wisconsin had weak laws regulating metallic mining. Churchill worried about a new mine proposed in his community by the Kennecott Copper Company. How was Kennecott to be regulated? The Citizen's Advisory Committee agreed there was a problem and decided to intervene.[10]

The Office of Wisconsin Public Intervener had been created in 1967 as was the Wisconsin Department of Natural Resources. Its mission was to ensure an adversary process during regulation of water resources. Public rights would ordinarily not be effectively represented unless the public intervener was vigilant. The public intervener should "watchdog" state agencies and could sue them or developers when public rights in the environment were endangered. "All state agency personnel were required to cooperate with the intervener, who had full legal power to present evidence, subpoena, submit briefs, and appeal all administrative and judicial rulings concerning environmental matters," noted historian Thomas Huffman. "The Wisconsin public intervener was the first environmental public-interest office proposed and instituted in the United States."

In 1976 Wisconsin had two new public interveners, Peter Peshek and Thomas Dawson. The Citizen's Advisory Committee

11

established priorities for both and appointed Peshek to intervene with the Kennecott mine.

In November of the same year, Kennecott Copper Company sought permits from Wisconsin to mine copper ore near Ladysmith in Rusk County. Roscoe Churchill and his wife, Evelyn, started opposing the Kennecott mine after they learned the corporation had purchased eleven neighboring farms.[11] Instead of vacationing in places like Palm Springs or Boca Raton, the dedicated Churchills toured old mining sites throughout North America, personally examining any damage inflicted by mining companies. As head of the Rusk County Citizen Action Group, Roscoe attempted to block the Ladysmith mine. He predicted gloom and doom for the area if it became reality, said one observer, "painting a picture so dismal that one got the impression Ladysmith would become a mining version of Chernobyl."

Nearly one thousand citizens in Rusk County signed petitions that declared it premature to approve Kennecott's permit applications. Spearheaded by the Churchills, the Rusk County Board voted not to grant a zoning permit for Kennecott's mine until Wisconsin developed adequate state laws to regulate mining.[12]

Meanwhile, Peter Peshek quickly agreed with the Churchills. "Wisconsin was not then in any position to determine intelligently whether, or under what conditions, Kennecott should be permitted to mine its Ladysmith ore body," Peshek recalled. "Wisconsin did not have a comprehensive and integrated regulatory scheme for copper and zinc mining." Before the judge acting as the hearing examiner, Peshek argued that Wisconsin was not ready to grant permits to mine. The weight of his argument coupled with the Rusk County decision not to grant zoning permits to Kennecott led the judge to dismiss Kennecott's application.

One of Kennecott's corporate vice presidents described the state in an internal letter as "the People's Republic of Wisconsin." The people had won, at least temporarily. "A handful of citizens beat a multinational corporation," Peshek gloated. "They whomped the living tar out of 'em. There were eight legal proceedings and the people won every one of 'em that ever ended. Just a glorious story - to knock off Kennecott without a regulatory structure."

"Getting into bed with environmentalists might rub raw with many of our colleagues," wrote a Kennecott official in an internal memo shortly after the judge's dismissal, "but in this day and age, I cannot recommend a better course of action for expedition of our project." Exxon learned that practical lesson well.[13]

After Kennecott was rebuffed, state officials adopted an innovative process to plug several gaping holes in Wisconsin's mining regulations. The legislature had established a Special Study Committee on Mining. After Exxon's announcement about the Crandon project, the legislature also created a Senate Select Committee on Mining Development. Governor Patrick Lucey established a third group to focus on mining, an interagency group known as the Economic Development Coordinating Committee (EDCC).

"By the fall of 1976, two special legislative committees were active, and the EDCC had initiated nearly a dozen topical, multi-agency study groups in its Mining Subcommittee," noted one authority. The state struggled to find the right course and decided to concentrate on taxation and environmental regulation.

After Exxon announced the Crandon discovery, one of their public affairs officials bluntly proclaimed that the ore body was "among the top ten massive sulfide deposits in North America." The announcement astonished Exxon's mining-industry competitors. "My God," said an exploration official at Toronto-based Noranda Mines, "that's like inviting the state tax people to pick our pockets."[14]

In 1977 after Exxon's public relations proclamation about the magnitude of the ore body, the state legislature levied a stiff net proceeds tax on metal mining operations, devising a progressive graduated rate schedule with the highest rate of twenty per cent for net proceeds over $30 million annually, the rate Exxon would have to pay.

Exxon perceived the new tax as punitive, and for the next four years lobbied to reduce it. In contrast, supporters of the new tax argued that the state's natural resources were finite. They hoped to prevent a repeat of what happened early in the century when lumber barons cleared acres of forest without paying any taxes. "What we're proposing is no less fair than asking a millionaire to pay more in taxes

13

than a man who earns $10,000 a year," said State Senator Henry Dorman.[15]

When the legislature established the net proceeds tax, it also created the Mining Investment and Local Impact Fund to receive the mining tax revenues and a board to distribute the money to nearby communities that could suffer social and economic impacts. Towns, villages, cities, counties, and tribal governments were eligible to receive payments. Municipalities could use the funds for police and fire protection, highway construction and repair, monitoring the environmental effects of mining, legal counsel and consultants, and schools.

Most of the mining laws passed in the legislature between 1977 and 1982 relied on a process of consensus. "Consensus" was a negotiating method aimed at developing a unified position among special interest groups directly affected by mining. Exxon's lawyer, James Derouin, described it as a process by which "reasonable people communicate with each other to reach reasonable solutions to impossibly complex problems."

Four key people involved in the process from the onset were Peter Peshek, the public intervener; Kathleen Falk, legal counsel for the Wisconsin Environmental Decade; Derouin, legal counsel for Exxon; and Kevin Lyons, the attorney for the Town of Nashville. Representatives from Kennecott and Inland Steel were also represented, and Don Zuidmiller, the attorney for the Town of Lincoln, subsequently joined the group. Roscoe Churchill attended for a while, but quit in disgust.[16]

The group's chairperson, Representative Mary Lou Munts, was a respected Democratic legislator from Madison who chaired several environmental committees and subcommittees in the State Assembly. She believed her role was critical. "A chair must be a catalyst, mediator, and manager and often must suspend judgment in searching for common ground in the group. If the chair becomes too partisan, there is no one to search for and develop the thread of consensus."

Acting as the facilitator, Munts continually pushed for agreement, even small agreements, and kept people from being "too greedy." "Because I went to a Quaker college and participated early

14

on in the League of Women Voters and other volunteer groups, I have looked at evolving the 'sense of the meeting,' or 'building consensus,' as intrinsically valuable," she said. "We were addressing generally uncharted territory, and attempting to see whether we could agree what needed to be done and whether we could agree how to do it."

"My use of the consensus method," Munts continued, "has been pragmatic in the sense that it suited my personality, and it worked for complex and difficult issues....As a group becomes infused with cooperative problem solving, solutions develop that could not have been conceived individually or negotiated bilaterally. In fact, there would not have been bills, or at least bills as comprehensive in most of the environmental areas tackled, if traditional legislative bill development by an agency or interest group had been used."[17]

Wisconsin's Environmental Decade saw the benefit of shaping and developing regulations that would protect the environment. "When you have political power to win on your own accord, then you don't need to negotiate with the other side," reflected Kathleen Falk about her role as their legal counsel. But environmentalists didn't have enough power. "When you don't have political power to win on your own accord, and the environment rarely has, the only way you can move forward is by negotiating."

The most active and dynamic participants were Peshek and Derouin. Consummate negotiators, they touted the consensus process in speeches throughout Wisconsin. Derouin, a former Peace Corps volunteer, reminded audiences of the alternative to consensus. "You can pay counsel now to practice preventive law and attempt to resolve conflicts ahead of time or you can pay counsel later when the case gets litigated or fought out to the bitter end in the legislature because there was no willingness to negotiate along the way." Derouin continued, "He who thinks that litigation or a bitter fight in the legislature is cheaper and quicker and will produce a better result is badly misleading himself. He who thinks that the only role that an attorney should play is to demand, befuddle, confuse and confound the person sitting on the opposite side of the table is not serving his client."[18]

James Derouin

The parties need to prioritize their concerns and communicate their bottom lines. "After all, parties to the process are involved because of their own perception of their self-interests - and they will remain involved only as long as it appears as if their legitimate 'bottom lines' can be accommodated," said Derouin.

In Crandon, Derouin met with the project manager of the proposed mine, examined reports, studied the mine design, the location of tailing ponds, and the timeline for permits. He discussed problems with Representative Mary Lou Munts, corporate executives in Houston, Exxon consultants, DNR officials, the Environmental Protection Agency (EPA), Zuidmiller and Lyons, Peshek and Falk, and the chairpersons of the Town Boards of Lincoln and Nashville where the minerals were located.

Exxon's faulty understanding of the consensus process sometimes frustrated Derouin. Exxon's project team in Crandon held a position not identical to that of the Exxon executives in Houston. Exxon's battery of attorneys had a position as did their regulatory and public affairs personnel. Exxon hired a consultant to inspect the report of their first consultant, then a third consultant to look at what the prior two had done, and then a fourth to review the previous three.

"Somehow you [had] to blend them altogether," Derouin recalled. "It was all a very slow, laborious and expensive process."[19]

Derouin negotiated with every party, tried to meet their concerns, but he insisted on something in return. "I got from them a commitment that if I made them happy that then...my project could go forward. I did not want to face an anti-mining argument after having negotiated all of these points."

Munts thought the consensus group was lucky to have Derouin. "He'd go back to his client people...and he'd say, 'Boy, have I got a deal for you.' And they'd look at it and say, 'That's a deal?'...He got some things accepted really through force of his own personality and his liking to reach agreements. I have to say he really liked environmental legislation. He liked the idea that he was putting something on the books that was sort of public-spirited and that his people could live with."

The consensus group believed that if they did not use mutual agreements to protect the environment, conditions would worsen and mining without adequate protection would occur. Good negotiations could substitute cooperation for contentiousness, rational discussion and compromise instead of power-politics. Participants thought the presumption for compromise was a major strength that counteracted extreme positions that had to be abandoned.[20]

It made sense for Exxon to participate in the negotiations. The company had watched Kennecott fail, and Wisconsin's outdated mining laws and regulations didn't provide certainty or direction for any corporate manager. Certainty was crucial to Exxon. "As long as the developers know the rules and can expect stability, they are willing to accept tough environmental legislation and rules that they probably could otherwise politically defeat," said Peshek. "Certainty to Exxon and other mining companies [meant] that the flow of corporate mining operations would not be compromised or needlessly delayed," said Michael Vaughan, one of several lawyers representing Exxon.

Group members had faith in each other. "I trusted those folks," said Kevin Lyons. "I respected their brains. I thought they were honest people. Those are the people I'm willing to attempt to negotiate with."[21]

The first major consensus victory in the legislature was the Metallic Mine Reclamation Act (1978). It established the regulatory ground rules for metallic mining in the state. The law provided for a single master hearing on the adequacy of the DNR's environmental impact statement and authorized the DNR to grant mining permits, approvals, and licenses. The law also set procedures for mine waste disposal sites and for reclamation. The act recognized that unregulated mining posed environmental risks (reclamation was critical), and if degradation of the environment was to occur, it had to be within reasonable limits if mining was to be acceptable.

Consequently the DNR was required to draw up new administrative rules. "The consensus policy actors viewed rule formulation as being as significant as legislation development," said Thomas Evans, geologist with Wisconsin's Geological and Natural History Survey. "The primary issue was the nature of regulatory protection of ground-water resources, specifically water quality, in the context of a metallic mining project."[22]

The Solid Waste Disposal Act (1978) mandated special rules for mine-waste disposal and established the Metallic Mining Council to advise the Department of Natural Resources on how to develop those rules. New rules were drafted over a period of two years. "The nine-member MMC was strongly influenced by consensus participants," observed Evans, "who accounted for four of the original nine individuals appointed to the Council."

A Long-Term Liability Act created a mining damage claim fund and applied the concept of strict liability to mining companies involved in metallic mineral development. Mining damage claims were to be brought before the Wisconsin Department of Industry, Labor and Human Relations.[23]

On the issue of management and protection of groundwater, some DNR staff members disagreed with the judgment of the consensus group. They wanted to continue a policy of absolute containment of any pollutants, known as the policy of "no detrimental effect." On the other hand, industry viewed the quality standard as vague and impractical. What did "no detrimental effect" actually mean? Even some environmentalists thought it was unenforceable.

The consensus group adopted a different regulatory approach to groundwater protection, a proposal calling for "numerical standards or specified maximum permissible concentrations of certain chemical species, consistent with the federally mandated primary and secondary drinking-water standards."

Evans thought the state was better off with the new groundwater standard. "I know we had a non-degradation policy. But it wasn't a standard," he said, because there was no specific numerical percentage to enforce. With the old policy, Evans said, "We had all kinds of industries discharge into streams. We had non-degradation at the same time we had paper mills." The virtues of the new standard were predictability and standardization. Advocates conceded that it allowed pollution, but argued it was not unacceptable pollution. The DNR adopted the numerical groundwater standards in 1982.

Among the new provisions was a controversial groundwater quality standard at a mining project that allowed a compliance boundary of 1,200 feet from a mining and waste disposal facility. Groundwater quality at the compliance boundary could not exceed the standard.[24]

"Who had the combination of money, expertise, and political sensitivity to the scope of environmental concerns to do a first draft of the rules?" Peshek later asked. "DNR clearly lacked all three prerequisites," he said, disparaging the DNR. "The strategy developed was for the most liberal of the mining companies, Exxon, to write the first draft, and the others involved would continually remind the company of their concerns and position."

The Wisconsin Legislature passed the consensus bills by overwhelming margins with little debate. Rarely did legislators question the substance of the bills; disagreements had been smoothed over earlier. State-agency review was virtually nonexistent, primarily because it generally wasn't sought. "There was little room and no need for any agency input or initiatives," noted Evans. "Even the legislature's own Council Mining Committee...appeared to relinquish the field of major policy development to the consensus process."[25]

Although not directly connected to the consensus process, James Klauser, a Madison lawyer and prominent Republican, was a key political advisor, strategist, and lobbyist for Exxon. While

Derouin focused on environmental regulations, Klauser took the lead in lowering taxes on mining companies. "The key to lobbying," Klauser said, was to communicate with everybody "tell a basic story. It's just a lot of work." He and his assistant made presentations to "almost every legislator." They met with Chambers of Commerce, newspaper editors, and unions. "It was a very very broad effort." Klauser outlined the damaging effect of the existing tax. "It was a simple message, done with a lot of data."

James Klauser

John Schmitt, President of the Wisconsin AFL-CIO, was an important contact, said Klauser, "but the steel workers were more important. Labor doesn't work top down. Labor works from its component members....Although John Schmitt was supportive; he couldn't have been supportive if the unions involved wouldn't have been supportive."

Klauser urged Exxon's Crandon project manager, Robert Russell, to become active in lobbying major interest groups, and Russell succeeded in winning the backing of the state teacher's union, the Wisconsin Education Association Council. The teacher's union, large in number and impact, became convinced Exxon's mine would bring jobs, tax revenue and would be environmentally safe.[26]

From 1979 to 1984 Klauser and Derouin participated in complex negotiations. In September 1980, Exxon cancelled a major phase of the Crandon project, a $30 million underground experiment of the ore body, a 10,000-ton pilot metallurgical test. The decision to cancel was carefully timed and politically motivated, the initial step in a campaign to lower the net proceeds tax. "They've decided it's time to push the [tax] issue," said Kip Cherry, mining specialist for the State Department of Development. "It's obviously their highest priority."

Klauser's lobbying worked. The legislature reduced the highest tax bracket from 20% to 15% and allowed significant deductions for extracting, milling, transporting, and smelting ore, administration and labor, supplies, repair and maintenance, depreciation, royalties and reclamation, reducing the total tax by 50%. The Wisconsin Department of Revenue developed a financial model with estimated revisions that would reduce Exxon's annual tax from $19.3 million to $9.3 million. The tax reduction greatly improved Exxon's economic prospect for the Crandon project.[27]

Derouin judged the new laws and rules excellent environmental legislation - balanced and understandable ("at least by lawyers'" standards), but others seriously doubted the merits of the consensus process and its resulting legislation. It did not incorporate the views of outsiders. Special interests dominated the consensus process, critics later said, and the legislature should have given the new laws and rules more than cursory review.

"The term consensus was badly misapplied to this process," said the DNR's Stan Druckenmiller. "The DNR's department staff wasn't there....The so-called extreme environmental groups were not involved....The tribes had little or no involvement." Druckenmiller thought it a "horrible mistake" to exclude the tribes.

The three tribes nearest the Crandon Mine were not at the consensus table, nor were they invited. "I, personally, did not have much interaction with the tribes," said the DNR's Linda Bochert. Why weren't the tribes involved? Representative Mary Lou Munts was asked. Her brief response was that tribal governments "have the independent nation psychology. It is hard for them to be part of a group to achieve consensus."[28] Munts' remarks, which may sound flippant and dismissive, may have reflected reality. At the time the

21

tribes were not interested in cooperating with state government, let alone the consensus process.

Roscoe Churchill initially took part in consensus meetings, but he quit after deciding that Exxon had set the agendas and believing mining would be allowed and water degraded. Under the revised rules, a mining company could degrade the groundwater to a national standard lower than that of drinking water in northern Wisconsin. Mark Patronsky, legal counsel to the Assembly Mining Council during the consensus process, observed: "I can assure you that all the parties to that discussion were looking over their shoulder at Roscoe Churchill, because he was asking an essentially valid question: Why should we do any damage to groundwater?"

Critics also challenged the "secrecy" of the process. "The people of the state are being cheated by this closed process of determining standards for water quality," said Churchill. Mining companies had the "first say-so about our groundwater quality before the public [had] a chance," an environmentalist added.[29]

"Press reports about the public hearings emphasized that the groundwater rules presented were a significant compromise between mining companies, state agencies and environmentalists," said mine critic Al Gedicks. "Very little coverage was given to the views of those who opposed the rules, despite the fact that the opposition constituted the majority of those testifying at the three DNR hearings held in various parts of the state. Furthermore, the people speaking in opposition were primarily from those communities that would be affected by mining operations."

Exxon and other special interests could afford the best lobbyists, attorneys and staff, and could make large campaign contributions to key political leaders. Environmentalists, Indian tribes, and citizens groups had few resources and limited influence in the legislature. In effect, consensus disenfranchised some, and it didn't include any group that was politically weak. "When an effort is made to keep the negotiations to a manageable size, it is those groups that are most likely to be excluded - despite the fact that their interests and views may be quite valid," said Douglas Amy, who studied the process. If a group posed no threat to the passage of a policy, it could easily be ignored.

"The main question of the local citizens and environmentalists - whether mining should in fact be taking place - was totally excluded from the agenda for discussions. But by assuming that mining would take place and negotiations would cover only the question of mitigation, the conveners were either consciously or unconsciously acting in favor of the mining companies....What is so politically problematic about the "consensus process" is the illusion it gives of always being an open, democratic, and fair process."[30]

Skeptical legislators felt helpless regarding bills recommended by the consensus group. Representative Thomas J. Crawford wrote to Roscoe Churchill about one bill. "As you know, it came out of one of those awful consensus processes, where industry attorneys are allowed to dictate the language of the bill....My amendments were routinely rejected by those supporting the consensus process." The proposed legislation "was printed and made available to the Legislature only a week before passage by the Assembly, [and] most members...had very little idea as to the details of the bill. The Department of Natural Resources was also absolutely no help on this bill at all." "While I am depressed and disgusted by the bill and the process used to put it together, there is nothing that I could do in the legislative process to change the result."[31]

Advocates of consensus praised the process for being "reasonable" and for avoiding "radical" positions. "At one time, abolitionists, women suffragettes, and civil rights marchers were seen as 'unreasonable' and 'radical,'" noted Amy. "But these groups and their positions now seem eminently reasonable."

Alluding to the many compromises made by the consensus process, a mine critic, George Rock, contended that "The best of consensus is going to give my kids only half of the environment."[32]

But critics of consensus may have been unrealistic and may have overstated their case. Would a better result have emerged had the state legislature performed its normal role, given the influence of the mining companies and business?

*   *   *

While Peshek and Derouin engaged in stimulating negotiations and helped enact new mining laws, the prospect of a giant mine next door bewildered the Indians at Mole Lake. They slowly learned mining could pollute Indian reservations and mining corporations would give little back to the tribes. They knew tribes had few allies, had limited funds for hiring lawyers and consultants, and had been ignored during the consensus process.

The first thing Sylvester Poler thought about when he learned that Exxon wanted to lease tribal land was his 20' by 20' garden. "I would have no protection that would prevent [them] from dropping a rig down in my garden."

"We were largely naive," recalled Poler. "Being naive meant we were ripe for exploitation." The Indians worried about being the victims of a "large disaster," and they didn't have the power to stop it.

Most of the proposed mine property was located on land owned by the Connor Lumber Company of Laona. After three years of lease agreements and negotiations, in 1978 Exxon agreed to purchase 1160 acres from Gordon P. Connor and to pay him royalty amounts based on the size and quality of the ore. Other smaller property owners reached a similar agreement with Exxon. The company had minimal trouble obtaining leases from Connor and from other white land owners near the mine, but negotiations with the Sokaogon Chippewa proved difficult.[33]

On September 15, 1976, Charles Arbaugh of Exxon presented Charles McGeshick, the Mole Lake tribal chairperson, with a check for $20,251 for the exclusive rights to prospect on the reservation. McGeshick apparently kept the check in his pocket for nearly a week before tribal leaders learned the deal surrendered mineral rights on the reservation to Exxon. The check was torn to pieces.

24

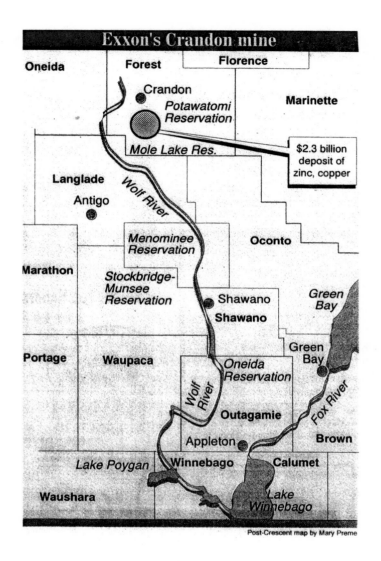

Exxon's Crandon mine

Oneida

Forest

Florence

Crandon

Potawatomi Reservation

Marinette

Mole Lake Res.

$2.3 billion deposit of zinc, copper

Langlade

Wolf River

Antigo

Menominee Reservation

Oconto

Marathon

Stockbridge-Munsee Reservation

Shawano

Shawano

Green Bay

Portage

Waupaca

Oneida Reservation

Green Bay

Wolf River

Outagamie

Fox River

Appleton

Brown

Lake Poygan

Winnebago

Calumet

Waushara

Lake Winnebago

Post-Crescent map by Mary Preme

Five weeks later, McGeshick and Arbaugh met again. In Exxon's version of the meeting, McGeshick insisted on a counter proposal of a $50,000 payment for two years along with a guarantee that if there was ore suitable for mining on the tribal land, "Exxon would conduct only subsurface mining." Exxon flatly rejected the offer and demanded return of the original check. (McGeshick must have told them it had been destroyed.)

"They offered us $20,000 for leasing and exploration of our land," McGeshick later explained. "We've sat down with them three times and I've talked to them on the telephone several times, but we're nowhere near an agreement."

In subsequent negotiations the Sokaogon tribal council demanded $20,000 the first year, $30,000 the second year, $4 million in advance royalties in the third year, and $2 million a year after that. Exxon rejected the proposal and negotiations ended.[34]

When visiting the reservation, some Exxon officials offended the tribe. "They called our wild rice a weed," Sylvester Poler recalled. "Something like that sticks in my mind, that they would call our wild rice a weed in front of us." It also bothered Mole Lake residents when Exxon posted "No Trespassing" signs on the proposed mining site where Indians had been accustomed to hunting and fishing.

After the unsuccessful negotiations, most of the Mole Lake tribal leaders and the tribal council refused to talk with Exxon officials. "We feared that by talking to them they would take that meeting, create a spin on it to say there was negotiations going on," explained Poler. The tribe did not want it perceived that they were negotiating.

The 1981 state law reducing the net proceeds tax on Exxon Minerals also included a provision that extended payments to Native American Indian communities that were affected by mining activities. The provision entitled the Sokaogon to $100,000 a year for up to four years during the mining permit process and additional funds after the start of construction.[35]

Gradually the Mole Lake tribe gained a better understanding of what was happening. In the process, most tribal members decided that preserving their culture and resources was more important than negotiating a financial deal with Exxon. When Americans for Indian Opportunity, a private foundation, held a conference at the Johnson Foundation headquarters in Racine, Wisconsin, Mole Lake sent a delegation. A major theme of the conference was damaging contracts with large corporations. "It was an eye-opening meeting, which made us think we'd better watch out for ourselves," said Dan Poler.

Poler asked for assistance from one of the conference speakers, Al Gedicks, a graduate student in Sociology at the University of Wisconsin-Madison, and founder-director of a public interest organization, the Center for Alternative Mining Development Policy. Gedicks agreed to help and secured several private and public grants for the tribe. "I wore out three different cars going from Madison to Mole Lake [and back]," said Gedicks. Mole Lake leaders were unanimous in their appreciation. "Al Gedicks has been with us from day one," said Sylvester Poler. "He was very loyal."[36]

The goal of a two-day conference at Nicolet College in Rhinelander, Wisconsin was developing strategy to stop mine development on Indian land. "The Mole Lake Chippewa obviously can't stop Exxon USA from developing its large zinc-copper lode at [Crandon], but the band attended the conference in force to hear some horror stories from Western tribesmen," one newspaper reported.

"What seemed to emerge from the many sessions at Nicolet College was a desire - really more a demand - that mine development shall not exploit Indians....Corporations intent on mineral development of Indian reservations polluted heavily and gave too little back to the tribes for the ores they got."

The Indians of Mole Lake wanted to know if valuable minerals did rest underneath their reservation. Why did Exxon want to lease it? The tribe convinced the United States Geological Survey to conduct an inventory of the reservation. Drilling showed that zinc and copper were indeed beneath reservation land. The tribe concealed the study's results in a garage. "Now we knew why they wanted to explore, get the lease, and get first development rights," said Sylvester Poler.[37]

In 1978 officials at the Bureau of Indian Affairs took a Mole Lake delegation, including Sylvester Poler, on a tour of a mining site in Leadville, Colorado. They were shocked to see a tailings pond that had spilled waste rock to a valley that "looked like a large cement lake." "We noticed a lack of wildlife," said Poler. "We noticed the company was taking the path of least resistance."

At the suggestion of Gedicks, the tribe contracted with a research team Gedicks had organized. Coact Research Inc. of Madison received a $70,000 grant from the U.S. Commerce

Department's Economic Development Administration to study the potential impact of the Exxon mine on the Sokaogon. Gedicks recalled:

"This group...helped document the hypocrisy of Exxon, which sought to project the public image of a socially and environmentally responsible corporation, by detailing Exxon's negative track record with other mining projects. This information provided the Sokaogon Chippewa with a way of gauging the reliability of Exxon's promises, based on the company's past performance record....The COACT study was designed to identify these project vulnerabilities so the Sokaogon could more effectively defend themselves against Exxon."

Most tribal leaders opposed the mine, but they worried about pro-mining people within their own community. "We knew that the mining company would try to divide [Mole Lake] against itself, create and stir up controversy," said Sylvester Poler. "We spoke vigorously against anybody who took a pro-mining position....We had to use the 'people power' within the community." Consequently tribal leaders ostracized pro-mining Indians.[38]

CHAPTER TWO

MAKING PROGRESS

As time went on, Exxon's public relations and lobbying improved dramatically. Their renewed efforts captured the outspoken support of business groups, especially in northern Wisconsin, and the Governor thought the company's plan had merits. The company stressed the economic advantages of the mine. Hundreds of jobs would be created; nearby communities would gain tax benefits.

As company engineers focused on protecting groundwater and surface water, two major elements loomed in the background. One was fluctuations in mineral prices. The other was dismal financial prospects for mining companies. (Exxon had consolidated its mineral divisions into a new company, Exxon Minerals, but the new firm was losing money.)

What impressed the general public about the Crandon project was its size. "Everything about it is big: big money, big minerals, big corporation, potentially big economic benefit or big environmental problem," observed a reporter for the Milwaukee Journal.

Crandon mine executives didn't know exactly how to gauge Wisconsin's Department of Natural Resources. The department was willing to negotiate and seemed to be making progress with its environmental studies, but it was slow and bureaucratic. To the DNR the Crandon mine presented an extraordinary challenge. Operating under new mining rules and pressured to expeditiously move on permitting, the department needed to regulate the huge project *and* protect natural resources and certainly prevent dangerous leakage from the mine's giant tailings ponds.

By mid-1977 Exxon's project team on the mining site began feasibility studies. The project manager in Crandon from 1977-1984 was Robert Russell, a native of North Carolina, who had worked as an exploration geologist in British Columbia, the Yukon Territory, and Alaska. The father of twelve children, the six youngest still at home, Russell built a six-bedroom house on the shores of Lake Julia outside of Rhinelander, and set out to manage a team that grew to thirty-eight people. Working for Exxon never strained his ethics. "I've never had any conflict with my beliefs and my work," he said.

"Because Exxon is a very good company to work for, I've never had to do anything I didn't believe in because of working here."[1]

Russell looked forward to directing the most advanced mine in the world. Far from despoiling the north woods, the mine would not damage the landscape at all. "That's definite," he said. At a mining conference in Denver, he had been "embarrassed" by the industry's attitude toward environmental protection. Clean air and water would cost too much. Russell felt differently.

Open meetings in Wisconsin did cause some concern. "We had to conduct these negotiations in a fish bowl," said Russell's top assistant, Barry Hansen, the project's permitting manager. Hansen never knew if a reporter was present. "You had to be very careful," said Hansen. "I was the guy in the middle. I was between our management in Houston [and the DNR]."

Friction arose between Exxon's project team in Crandon and headquarters personnel in Houston. "Headquarters people, who weren't involved on a day to day basis, wanted to review every criteria," said Hansen. Houston didn't understand Wisconsin - the attitudes and the concerns of people - and what really needed to be addressed. Waste disposal was an example. "We proposed…a double lining at the bottom of the tailings pond," said Hanson. "That was unheard of." Engineers in Houston "thought we were giving away the farm." A double lining was considered too expensive and unnecessary. Houston thought the Crandon team was making more concessions than necessary.[2]

In spite of problems with Houston, Exxon's public relations in Wisconsin improved dramatically from 1977-1981. Leading the public relations strategy was James Klauser. "Klauser was the master of political strategy in a positive sense," said Russell. Klauser insisted that Exxon be a "positive influence in Wisconsin," "the positive story had to be told" in an "enlightened, environmental manner."

Lobbying in the state was enhanced by the interests of Wisconsin's manufacturers of mining and construction equipment, companies like Allis-Chalmers Manufacturing Company, Bucyrus-Erie, and Harnischfeger Corporation. "If mining takes hold in Wisconsin, an estimated $1.5 billion could be spent on site development over the next several decades," wrote Business Week in

1977, "and that could mean equipment purchases of more than $350 million from Wisconsin manufacturers."[3]

The mine also gained endorsements from major northern Wisconsin business groups. The Crandon Area Chamber of Commerce championed the project, anticipating hundreds of new jobs. Many Rhinelander businesses joined supporters. "If it goes and they create 1,600 jobs, it will mean millions of dollars to the north woods area," said Bill Hyland, executive secretary of the Rhinelander Area Chamber of Commerce. Hyland said the Chamber supported Exxon "100%."

Public opinion in the Crandon area favored the mine, but there were some reservations. "The relative level of acceptance was due in part to Exxon's masterful public relations job," wrote the Milwaukee Journal. "Company officials have attended literally hundreds of meetings with citizens to talk about their plans and answer questions."[4] Exxon had also distributed thousands of multi-colored booklets with photos, maps, and diagrams of the mine.

"Meetings, meetings, meetings - it seems to be a Wisconsin way of life," said Russell. "Our approach has been one of being as communicative as we could relative to our plans. It's been a very open dialogue, and we've tried to listen very closely." Local officials appreciated the access. "I can call them on the phone and ask any question I want and they'll answer me," said Crandon Mayor Gwinn Johnson. "They've been very cooperative, very cooperative."

But folks in the towns of Nashville and Lincoln worried. While people in Antigo and Rhinelander could talk excitedly about hundreds of jobs, "They would feel differently if they had a tailings pond in their back yard," said John Schallock, chairman of the Town of Nashville.

Democratic Governor Tony Earl, elected in 1982, told officials of the Exxon Minerals Company that he supported their efforts to open the mine. In early August 1983, he met for two hours with Exxon officials in New York and told Exxon the state would help to clear the paperwork as quickly as possible. "You know I'm not a pushover," Earl said he told Exxon executives. "But I want you to know I'm not intransigent."

Some Wisconsinites believed mining an unacceptable undertaking. "I don't believe that," said Governor Earl, a former secretary of the Department of Natural Resources. He believed mining could be done without ruining the environment. Earl criticized those who automatically opposed mining. "There's something wrong with the philosophy that says it's wrong to mine metallic minerals, but you can put a gravel pit anywhere and just leave it.

"I'd rather the mining was promoted by Exxon than by some fly-by-night company that would cut corners at every turn," he said. "It hasn't been like pulling teeth to get cooperation from Exxon." Earl urged state agencies to swiftly process non-controversial permits as a "ministerial function" during the next few years, even though major policy permits on Exxon's mining project would take time.[5]

The primary focus in the company's public relations was spotlighting the project's economic benefits. The Wisconsin Department of Revenue conducted computer projections that predicted Exxon would net $973 million over a twenty to thirty year life of the mine. There were huge economic advantages for the state as well. Using 1984 dollars, not counting inflation, the department forecast that Exxon would pay $117 million in net proceeds taxes and another $95 million in state corporate income taxes.

Some state officials thought Wisconsin's business reputation depended on the Crandon project. "It's a question you have to ask when you put in any industry," said Robert Ramharter, the DNR's Crandon project coordinator. "The answer is if mining is going to be a social or economic advantage to have, with proper environmental safeguards, then it's desirable."

In presentations to community groups and in its newsletter, Crandon Report, Exxon company officials stressed an optimistic economic forecast if the mine won approval. It would create many new jobs, reduce unemployment, increase personal income, greatly expand the local property tax base, and increase business activity in adjacent areas.

One thousand jobs would be created during the four-year construction period. Half of those jobs could be filled by "qualified" local residents. At full production, the complex would employ about 800 people, half of whom would be hired from the local area.

"There's a feeling this will be very good for the county," said Erhard Huettl, Forest County Board Chairman. "The average common wage around here is about $3.40 an hour. It's damn hard to own a car, buy food and shoes for your children on $3.40 an hour."[6]

Crandon, Antigo, and Rhinelander would experience increases in retail trade, augmented banking and insurance activity, and other improvements, the company contended. Employees would spend more than $300 million on goods and services in Wisconsin during the construction phase and $40 million more each year of operations.

Most of the mine facilities would be constructed in the Town of Lincoln and would multiply the property tax base of the town approximately ten times. The remaining construction would be in the Town of Nashville and would double Nashville's property tax base. The mine would also double the property tax base in Forest County and triple the Crandon school district's property tax base. Native Americans would benefit by acquiring mining jobs and by receiving tax revenues paid to towns, the county, and the Crandon school district; mandatory payments could provide "seed money" for Native American enterprises.[7]

In December 1982 the company submitted its Environmental Impact Statement (EIS) to the Department of Natural Resources. It described what the Crandon Mining Co. proposed to build. The company would evaluate alternatives before selecting mining, milling, and related technologies; produce baseline studies of the ecological conditions at the site; and project the effects of construction and operation of the mine and the mill.

The EIS submittal triggered a specific sequence of steps. The DNR had to review and verify the information in the EIS, and after completing that study, prepare and issue a Draft Environmental Impact Statement (DEIS). Following a public hearing, the DEIS would be revised, and a Final Environmental Impact Statement (FEIS) would be prepared and issued. Finally, a master hearing would be held on the FEIS, complete with sworn testimony.[8]

*   *   *

Daunting engineering problems faced the company. Barry Hansen told his team that discussion of sites or location was premature. "We are all going to talk about criteria. What is the most important thing for us to do?" Discussion created consensus that the first priority was protecting groundwater and the second was protecting surface water.

The surface mine facility would include a head frame, an apparatus for ore hoisting, and a milling complex. The complex reduced the size of the ore and concentrated the minerals so they could be shipped out of state for additional refining.

The underground mine would use the "blast hole open stoping" method for extracting the ore. After workers placed explosive charges in holes drilled vertically through the ore, the charges would be detonated from the bottom to the top, breaking loose blocks of ore. The ore would then be taken to the main shaft, crushed and hoisted to the surface for additional processing.

The company planned to monitor water levels regularly in nearby lakes and wells because water would have to be pumped from the mine, possibly drawing down the groundwater over the ore body. "Despite our best efforts at avoiding and controlling water underground, it is not possible to guarantee 100 percent success," said Hansen in 1985. The mine's operation would affect some wells in the area, but Exxon Minerals guaranteed in writing to the nearby Towns of Nashville and Lincoln and to individual property owners in the area that the company intended to protect their water supplies. If the mine impacted water, the company would provide an equivalent source of water.[9]

"At this time there are about 25 private wells not owned by Exxon in the expected zone of influence," claimed the Crandon Report. "Most of these wells serve seasonal residences. *We are committed to making sure that neither the quality nor quantity of any permanent or seasonal resident's water is adversely affected by the development of the Crandon mine.*"

Most of the lakes in the mine area were fed from surface runoff and were *"perched"* well above the groundwater table, the company contended. The drawdown in the aquifer would not affect those lakes. The perched lakes included Oak Lake, Little Sand Lake,

Deep Hole Lake, and Duck Lake. "We and our environmental consultants expect no measurable effect on the level of these lakes during mining operations."

Spring fed lakes - Rolling Stone Lake, Ground Hemlock Lake, and Rice Lake - were connected either directly or through surface waters to the groundwater aquifer, but there was no need to worry because all were well outside any significant area of draw down. "Because of their distance from the ore body, no effect on the levels of these lakes is anticipated." There would be a reduction of flow to Hemlock and Swamp Creeks, but Hansen claimed that the company's extensive hydrologic investigations proved that the reduction would be small and have no effect on the area's wildlife and fisheries.

The company planned to design a sophisticated water treatment plant applying the principle of reverse osmosis. "By employing reverse osmosis, high quality process water is produced which will be suitable for re-circulation to the mill," said Hansen. "The Crandon concentrator will be the first one of its type in North America to operate on 100% recycled water." A controversial feature of the mine process involved wastewater disposal. Water pumped from the mine and not needed for the mine's operations would be treated, carried in a pipeline and discharged into Swamp Creek, a tributary of the Wolf River.[10]

The most environmentally sensitive component was the tailings pond facility and system. Tailings, a mixture of pyrite and rocks of volcanic origin, had no commercial value, but when oxidized and exposed to water, they form exceptionally dangerous sulfuric acid. The plan was to return about half of the tailings underground to fill the voids in the mine created by the mining process. The other half would be disposed of on the surface. Each day the company would pump about 4,300 tons of tailings to the mine waste disposal facility in the form of a slurry, a mixture of water with the fine-grained ground waste rock tailings. At the mine waste disposal facility, the company explained, the slurry would be discharged into a specially prepared pond of about 100 acres. "The first pond would last about five years and, as it neared capacity, a second pond would be constructed, and ultimately a third and fourth."

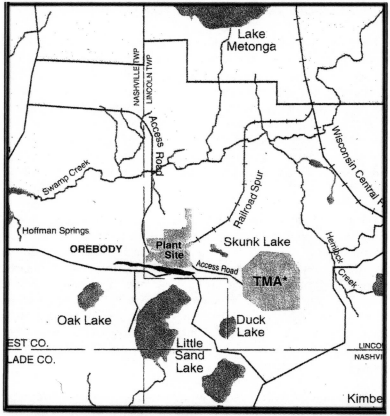

Tailings Management Area

After the company filled the first pond, reclamation work would begin immediately while it filled the second. "At the end of the mine's life, the final pond would be sealed, covered and re-vegetated. This multi-staging of ponds minimized the amount of land disturbed at any given time and provided the opportunity for early reclamation." It allowed the company to incorporate new technology as the second, third, and fourth ponds were developed.

The design of the liner assembly of the waste disposal facility would include a continuous liner on the bottom, sides and top of the pond. A drain would prevent the ponding of water beneath the wastes that might otherwise put additional pressure on the liner. "This ponded water is withdrawn and pumped back into the plant's circuit.

The result of this drain and liner assembly is a very low level of seepage from the bottom of the pond."

The top cover designed for the tailings pond consisted of a seal of bentonite soil admixture. "On top of that there will be a 40 mil polyethylene membrane," said Hansen. "On top of that will be a drain layer which will help draw water away from the top of the waste." The whole assembly would be covered with at least five feet of glacial till and topsoil. "In all aspects of the design, special care has been given so there is minimal change in the north woods environment surrounding the proposed mine site," claimed the company's newsletter.

The project team labored long hours. "I remember working fourteen hour days," said Edie Franson, a secretary in the mining office, who also worked some Saturdays and Sundays.[11] As Robert Russell and his team labored in Crandon, broader mining issues were engulfing Exxon and the entire mining industry. In 1980 Exxon had consolidated its various minerals divisions into a new company, Exxon Minerals Company. Exxon Minerals suffered losses of $97 million in 1981, and $114 million in 1982. The Wall Street Journal reported in August 1982, that Exxon had entered the mining business as a logical extension of the oil business. Unlike other oil companies, it had "eschewed very large acquisitions...and relied mostly on its own exploration and development." Exxon Minerals President, William M. McCardell, thought this approach was "a long, hard way to do it," and that the division had "suffered from 'birth pains' as well as from low metal prices." Employment at Exxon Minerals dropped from 3,000 in mid-1981 to slightly more than 2,000 a year later. The Journal reported that Exxon might pull out of the minerals business altogether unless the new company improved.

In December 1984 Business Week forecast gloomy times for mining, an industry that once had a proud heritage. "In their day, North American mining companies ranked among the world's industrial elite. Amax, Anaconda, Asarco, Kennecott, or their forerunners helped settle the West. Huge family fortunes derived from mining, including the Hearsts' and the Guggenheims'. Mining companies were among the first U.S. multinationals dominating world markets during the boom that followed World War II."

Now after a long period of painful losses, mining companies were reeling from the chronic problems of foreign competition, shrinking markets, huge debt, and depressed prices. "Three or four major metals producers may even be forced out of the business over the next few years. In a very real sense the industry is dying." Vulnerable to fundamental forces, "It is an industrial activity in which, these days, the developing nations have an almost unbeatable pair of economic advantages: cheap labor plus very-low-cost reserves."

The oil companies were having a problem making their new investment pay off, The Wall Street Journal reported. The bright profit prospects "suddenly look lackluster at best." The chief executive at one established mining company relished Exxon's predicament. "The oil boys are learning some of their grandiose mining plans just can't be achieved."[12]

Officials at Exxon's headquarters in Houston continued to question the painstaking process necessary in Wisconsin to build an environmentally safe mine. By 1984 corporate decision-makers felt Robert Russell was not a disciplined manager who could meet deadlines or stay within budget. "My judgment at the time was that they were wrong," said J. Wiley Bragg, Russell's primary supporter in Houston. "They didn't understand...the imponderables. Russell was doing everything he could to meet their requirements, [while] recognizing the realities of being onsite." Russell felt micromanaged. "Whereas people had left me alone for all the years that we did all the good work in developing the laws and regulations and changing the tax law, now I was going to get 'special attention' from Houston. The only defender I had [in Houston] was Wiley Bragg."

By mutual agreement, Russell left his Crandon position in 1984 and took an assignment in Africa. He believed "Exxon was refocusing on energy and de-emphasing metallic minerals. They were not going to go anywhere with this project."[13]

Donald B. Achttien replaced Russell, becoming general manager of the Crandon project in mid-February of 1984. Achttien (pronounced Octane) did not have a flair for community outreach and was not as "friendly and cordial" as Russell, in Bragg's opinion.

In May 1985, Exxon Minerals announced a major change in the Crandon mine plans. Because of depressed minerals prices, it intended to improve the project's economics. The company now intended to first mine the zinc portion of the ore body and delay development of the copper portion. The new proposal would lower investment costs, reduce required manpower from 800 to 600, extend the life of the mine, and reduce the construction period from forty-two months to about thirty months. Unfortunately, the change also extended the permitting process by one more year, and that process already seemed like an eternity.[14]

*   *   *

"By 1990, forty-five states had centralized their environmental administration in some kind of consolidated department," observed Thomas Huffman. Wisconsin did so in July 1968 when it established the state's Department of Natural Resources (DNR). It soon became the dominant environmental institution in the state and the first environmental "superagency" in the nation. By early 1970 it had two thousand employees and a biennial budget of $67 million. The DNR performed valuable environmental work, but it was also viewed as aggressive, impersonal, and bureaucratic. The specific, deeper grievances usually involved economic factors. The DNR had the responsibility for protecting the state's land, water, wildlife, and other natural resources, and its regulating activities angered many. As William Thompson noted, "Private businesses, public utilities, and local governments wished to continue dumping their waste into Wisconsin's rivers and streams. Hunters and fishermen wished to pursue their sports unencumbered by a growing host of fees, tags, permits, and seasonal and geographical regulations. Farmers wished to plow, plant, and graze their livestock as they pleased, without regard to the purported needs of ruffed grouse, white-tailed deer, or brook trout."

In mid-June of 1976 when DNR first met with Exxon to discuss the kinds of information needed to review the mine project, they had no specific mining unit that could study a project as gigantic as Crandon. "We didn't have any mechanism to coordinate our internal reviews," said Robert Ramharter who became the DNR's coordinator of the Crandon project. When Ramharter requested the assistance of a hydrologist within the department, the answer he received was, "We don't have anybody."

Gradually the situation improved and staff was added. "We do ten times what most states do in terms of getting involved in the project, finding out what they're doing, verifying the data," said Ramharter. James Derouin agreed and said, "I would like anyone to go to any other state or to the federal level and match their rules against what we have in our state. Mining, if and when it occurs, is going to be linked to the highest environmental standards."

Planning for the mine inched forward under the weight of Wisconsin's new laws and rules. The DNR was constrained by those laws and rules, a constraint unappreciated by many opponents of the mine. DNR didn't have the authority to decide if mining was good or bad; the legislature had already decided that mining was beneficial. All the DNR could do was make certain that the mine conformed to the new laws and rules.[15] The DNR required massive environmental impact reports, social and economic studies, and payment by Exxon of the state's expenses accrued when processing permits and research.

By mid-1977 Exxon's baseline studies were underway. The following year Exxon hired consultants to conduct socio-economic research, projecting how mining would impact neighboring municipalities, Native Americans, demographics, economics, and public services (schools, roads, fire protection, and others). The DNR then hired consultants to help its staff evaluate Exxon's analysis. The DNR needed to conduct an independent evaluation of the proposal, to assure its FEIS was accurate to a "reasonable degree of scientific certainty."

The burden of proof for the proposal meeting Wisconsin's standards fell on Exxon Minerals. "If our verification demonstrates that Exxon's information is accurate, we will rely on it," explained a DNR directive. "If we determine that the information is incomplete or inaccurate, we will request additional material from the applicant or develop it ourselves as appropriate."[16]

Wisconsin law prohibited a mining company from causing an "unreasonable detriment to public rights" in surface waters. Public rights included fishing, swimming, boating, and aesthetics. "Permits for the project cannot be approved if the analyses show that significant impacts to surface waters would be expected and would not be repaired, or 'mitigated,'" said the DNR.

Initially the DNR's work focused on verifying Exxon's baseline studies. State law required the company to document existing environmental conditions at the mine site before mining began. These included surface and groundwater quality, air quality, geology, wildlife use, fish life, plant life, forestry and wetlands.

DNR officials scrutinized Exxon's plans, oversaw the company's sampling techniques, gathered independent information, and inspected the accuracy of the company's laboratory techniques. They split samples with company engineers and scientists and analyzed them on their own. As Exxon sampled the water quality of lakes, streams and rivers, the DNR staff observed the handling of aquatic samples. "To assess water quality, investigators measured drainage patterns and physical qualities of water like temperature, clarity, flow and speed," said a DNR report. "They also chemically analyze the water to determine acidity, hardness, amount of dissolved oxygen, amount of organic nutrients and other constituents that pinpoint what kinds of plants and animals can live in the water."[17]

In May 1983, while characterizing Exxon Minerals' report on the mine as a "major step forward," the DNR criticized the report's shortcomings. In a four page letter and a fifty-two page point-by-point analysis, the DNR told Exxon, in effect, that the process of getting all the necessary licenses and approvals for its mine would not be easy.

"Exxon seems to dismiss many of the Department's comments regarding monitoring," the DNR's Larry Lynch impatiently wrote his colleague, Gordon Reinke. "The comments were not merely suggestions on the Department's part. Rather, Exxon should meet with the various Department programs to finalize the monitoring requirements." Lynch gave examples:

More information is needed concerning leak detection for the tailings pipeline. Where along the line will flow rate monitoring devices and pressure gauges be located? Approximately what magnitude leaks are actually detectable using the proposed system? How is it determined where a failure (total and partial) has occurred and hence where to initiate clean-up measures?

The DNR repeatedly questioned the design plans for the tailings ponds which the company said would hold about 31 million cubic yards of highly sulfuric mine wastes. Richard Schuff, chief of the DNR Bureau of Solid Waste Management, told the Crandon project team that it needed more supporting evidence to assure the DNR that groundwater in the area would be protected from heavy metal and sulfuric acid contamination.

"It's kind of a good news-bad news situation," said Barry Hansen about the tailings pond sites. "There is a sense of disappointment," he added. "The issues were thoroughly discussed, we thought. Apparently they are not understood as we thought they were." The DNR also kept questioning the depth of the proposed clay liner and the ability to keep water and air out of the tailings after the company closed the site.[18]

"The company was always reluctant to provide us with information, to provide us with resources," DNR's Ramharter recalled. "It was like pulling teeth to get them to do things. They were always pushing to move the schedule ahead, and we were always asserting that we did not have sufficient information and documentation or studies to proceed with the [DEIS]."

"It was difficult to get decisions out of them," agreed Stan Druckenmiller. "There was a level of arrogance on the part of Exxon that completely surprised us. They were very reluctant to give us information. We would send them a letter for them to convey some information, and they would send us a letter back saying, 'Here's some of it but we don't think you need this other stuff.'"[19]

The company had difficulty with proper modeling for the drawdown of water from lakes and streams. Drawing down water was unacceptable to the DNR because it disrupted fisheries, the flow of water, the rice beds, and public rights generally. "We weren't satisfied with what they were projecting in terms of the drawdown," said Ramharter. The company "didn't think it was much of an issue."

Druckenmiller had to remind others there was nothing predetermined about the Crandon project. He told Kip Cherry of the Department of Business Development that some terminology Cherry had recently used implied a certain inevitability of mine approval. "Since many regulatory hurdles must be crossed before [the project]...becomes a reality, we prefer the use of 'would,' 'could,' 'may,' over the currently used 'will' in describing mine development plans."

In 1983 a core staff of thirty DNR personnel worked exclusively on the Crandon project, and another thirty worked part time. The department's staff devoted 16,000 hours to the project in the third quarter of 1983.[20]

In November 1983, Ramharter said the department had been under "no pressure" from Governor Earl's office "to hurry studies along, cut corners or do a halfway job." However two decades later Ramharter admitted he had been under intense pressure to expedite permitting. "The pressures that I had were to get the project done...get the [FEIS] finished" The pressure "came from our top management in the [DNR]," mainly from an executive assistant.

Druckenmiller thought DNR's pressure was normal and legitimate, "but it is hard to agree that that was unfair. Is ten years long enough to finally get it into the public domain?"[21]

The company and its supporters agreed Wisconsin's permitting process was slow and lengthy. "Before coming here to speak, I looked at the articles written back then, and I was amazed to see those clippings identified the same issues we're struggling with ten years later," said Exxon's Don Achttien before the Wisconsin Newspaper Association in 1985. "We realized ten years ago that all of the specific questions and concerns wouldn't be put to rest quickly. But the dialogue has been going on a long time. I believe it is in everyone's best interest to bring the process to a point where a decision can be made."

A positive decision would not only benefit Wisconsin economically, it might help improve the state's business image, he said. "Simplification of the permitting procedure, granting of Exxon's mining permits, and the introduction of a new type of development activity by a major corporation would send positive signals to others." The message would be clear. Wisconsin was a good place to do business. "It's a good place to live, work and invest."

State Senator Lloyd H. Kincaid, a Democrat from Crandon and an avid mine supporter, was less diplomatic. In his view, the DNR's staff was a "bunch of left-wing flakes" who engaged in inexcusable delay in reviewing the proposed mine. "At a time when businesses are leaving the state, we should be more receptive to one that is begging to come in. To date, the DNR has not been receptive, but dilatory," Kincaid wrote the DNR. "From the time the Crandon project began and the DNR began its review, three governors have been in office," Kincaid added. "I believe it is time for your agency to show a greater sense of concern over the delays in decision-making on this important matter."

While Achttien and Kincaid articulated the pro-mine position, an alternate view of the mine was slowly emerging.[22]

CHAPTER THREE

"ALL TAILINGS PONDS LEAK"

An official in the Department of Natural Resources told a journalist there were few environmentalists, "except bird watchers," in northern Wisconsin. Bird watchers were okay, but they weren't the kind of people who would take on Exxon Minerals' zinc-copper mine. "Journalists who covered mining issues since Exxon announced its major ore discovery in 1975 have often noted a lack of active...opposition to the mine," agreed one newspaper.

The two observations were somewhat exaggerated and premature. There were many environmentalists in northern Wisconsin, but they had been quiet and inactive until a giant corporation in their midst, threatening their lakes and rivers, aroused resourceful and tenacious opposition. That opposition, lead by four individuals, an improbable congregation of nuns, and the state public intervener, while articulate and forceful, was not strong enough to stop the seemingly inexorable forces in favor of the mine. Only a bizarre decision by Exxon ended the permitting process in 1986.[1]

Critics contended that most of the rock that came from the Crandon mine would end up as tons of waste material, finely ground iron pyrites that are extremely dangerous, forming sulfuric acid when exposed to air and water.

Exxon claimed its tailings ponds would safely contain the toxic waste almost indefinitely, but their financial liability for the tailings ponds would end thirty years after the mine closed. All tailings ponds eventually leak, critics insisted. According to the Environmental Protection Agency (EPA), "the regulation of hazardous waste...must proceed from the assumption that migration of hazardous wastes and their constituents and by-products from a land disposal facility will inevitably occur." The Crandon tailings ponds, opponents feared, would leak hazardous waste into Swamp Creek and into the Wolf River. That tailings ponds leak became a mantra for the anti-mining movement, repeated thousands of times: ALL TAILINGS PONDS LEAK.[2]

45

In addition, constant pumping of water from the Crandon mine would lower the water table. Little Sand Lake could be totally drained in 7.4 years of mining activity, and other lakes would be similarly affected.

Mine critics also cautioned that Exxon may not pay any of the mining taxes it touted. Only those operations that showed a profit were liable for the net proceeds tax, and the entire United States mining industry was declining. If Exxon Minerals didn't make a profit, it would not have to pay any net proceeds tax. After all, the company had operated at a loss some years.

Opponents were also wary of Exxon's pledge to hire local workers and doubted projections that the population growth in the region would not overwhelm local communities.

"I also fear that once Exxon develops, there will be other mines all over," said one critic. "We'll end up like northern Minnesota or northern Michigan. That will be the end of northern Wisconsin as we know it."[3]

Four leaders emerged as major critics of the mine. One was George Rock. In 1983 at an anti-mining forum in the Nashville Town Hall, more than sixty people heard Roscoe Churchill describe mining as a danger to the environment and the social structure of northern Wisconsin. "We were new to the game," said attendee George Rock, and Churchill's message galvanized his listeners. "You guys must understand what is going on," Churchill challenged. "Get organized!"

Rock's father had been a teacher and conservationist who hunted and fished on the Wolf River. His son, George, an engineer, worked for thirty-seven years in the Wisconsin Department of Transportation in Green Bay, and on weekends he journeyed to his cottage on Pickerel Lake near the Wolf River. "I'm concerned," he said about the Crandon mine. "Very concerned." Diligent and tenacious, Rock had little experience as a public speaker, but the mining issue forced him to become one.

Exxon was irresponsible in its mining activities, Rock charged, pointing to a newspaper clipping about Exxon's failed shale oil project in Colorado where thousands of workers were thrown out of work.[4] Rock reminded citizens that for fifty years Forest County had promoted itself as a recreation area. The county published a map of Forest County, *The 4 Season Playground*. "If you look at this map closely in large red letters you will see a paragraph entitled 'Forest County, the Unspoiled North,' and below this heading the following invitation - 'The Forest County Board of Supervisors, the businessmen and resort keepers and the residents of this unspoiled vacation paradise cordially invite you to spend your next vacation in Forest County. This is an almost virgin, unexploited area. Its many lakes and streams provide uncrowded relaxation.' Note the reference to 'almost virgin, unexploited area,'" Rock concluded. "By definition, mining is exploitation at its worst!" Leading the Concerned Nashville Taxpayers, Rock helped organize a drive that recruited 291 residents and cottage owners to write town officials calling for a moratorium on mining.[5]

Rock flooded the offices of newspaper editors with letters and documents. "You have supplied me with a veritable ton of information," complained Mike Walter, editor of the Appleton <u>Post-Crescent</u>. "I have neither the room in the newspaper nor the interest in publishing so much, seeing as how it is to be done in reply to only one editorial....I do not want reams of background material for my files."

"I hope you weren't discouraged by what you thought was a small turn out Sunday," Evelyn Churchill wrote Rock. "You can bet that those there were anxious to learn and determined to do whatever they could....<u>Education</u> is the thing and you have been doing a great job of it."[6]

Herb Buettner operated two motels, a supper club, and a large raft rental firm along Highway 55, the Wolf River corridor. He was in constant demand by news organizations throughout the United States as an expert on the Wolf River. His family held deep environmental convictions. His oldest brother was a DNR forest ranger; another brother opened a fly-fishing shop in the late 1960s. "Between the two of them, they taught me conservation," Buettner recalled.

Back in the early 1960s, developers obtained a state permit to build a dam on the upper Wolf River that would flood a low, marshy area along the river to create a 1,500-acre lake for development. Buettner did research, hired a surveyor, and demonstrated that the dam would starve the river of spring water. He brought in state legislators to explain his position, then testified before the legislature. By a unanimous vote, the state legislature outlawed the construction of any dam on the Wolf River.

Buettner realized the Crandon mine could boost the local economy, but he distrusted Exxon's "sales presentation." "There were too many steak dinners with town officials for his taste." Exxon's "big-city" approach to courting local decision-makers didn't sit well with him.[7]

Reporters often located him in the Wolf River in waders, "hiking up a steep, forested ridge, fly rod in hand." A founder of the first Trout Unlimited chapter in Wisconsin, Buettner had many friends in the fishing and hunting communities. "To protect the waters of the area we ask that the proposed wastewater discharge pipe to Swamp Creek, a feeder to the Wolf River, be eliminated and seepage ponds be installed," said Buettner. "We ask that all chemicals used in the operation be recaptured and returned to the supplier for recycling or proper disposal. And we believe the tailings themselves contain toxic chemicals and must be permanently contained. Those which cannot be put back into the depleted mine should be cast into concrete or a better method developed to assure forever they will be no threat, or a perpetual containment expense for the County or State."

Buettner was the catalyst behind a petition in which visitors to the area demanded that ore remain in the ground if Exxon and the state couldn't assure protection for the Wolf River. "We must not allow politicians to compromise or trade off the clean waters of the Wolf River for political or corporate gain....Whatever happened to the moral concept that if it's clean, keep it clean?"

Buettner believed in the environmental principle: use, pass on, but don't diminish natural resources. Like George Rock, he found the Indian philosophy compelling. Mother Earth gives life, food and water. We have the responsibility to our children and grandchildren, seven generations hence, to protect it. "It's the only philosophy that makes sense," he said.[8]

Born in 1940 of Polish ancestry, Sonny Wreczycki and his family spent World War II in a German concentration camp. After enduring five more years in a post-war displacement camp,

Wreczycki arrived in the United States in 1949 and moved to Milwaukee in 1955 where he worked for the Reynolds Aluminum Company.

Sonny Wreczycki   Mimi Wreczycki

On most weekends Wreczycki traveled north to his cabin on the sixty acres of land he owned on Rolling Stone Lake in the Town of Ainsworth, a few miles from the proposed mine site. He attended his first meeting about the mine in 1976 and remained in the anti-mining movement for the next twenty-seven years. As president of Ainsworth's Mining Impact Committee, Wreczycki criticized Exxon's estimates of the benign effects the mine would have on nearby lakes. "Did they come in here and do a proper study?" he asked. "No. Instead they cherry-picked data...put it together to make the best possible picture for themselves."

He drew a glass of water from his tap for a reporter. "Take a taste of this!" he said. "Now that's good water, and it comes straight out of the ground." If the DNR permitted the mine, Wreczycki said, no one would want to drink from his well again.

Wreczycki's research revealed the <u>alternate</u> tailings pond site would be located on the west end of Rolling Stone Lake. Would the company and the DNR decide to use the alternate site without even consulting nearby residents?[9]

Wreczycki monitored mineral prices worldwide and studied company reports, finding flaws and inconsistencies. "They had creeks running the wrong way," he found, and had "swamps at the wrong locations." At a meeting in the Town Hall in Nashville, the Crandon Mine Company's Don Achttien said to him, "We spend all this money on studies, and you use them against us."

Wreczycki did impress journalists and some others with his research. "We continue to be impressed by your efforts in reviewing the documents relating to the Exxon proposal and by your knowledge of the local surface water systems," a DNR official wrote him. "In addition, you have raised a number of important concerns that need to be addressed as we evaluate the impacts of the proposed tailings disposal area."[10]

One of the earliest opponents of the mine project and the most radical was Al Gedicks. Gedicks "was a voice of opposition," noted Peter Peshek, a "loud voice." Al Gedicks "never saw a mine he liked," said a DNR official. Gedicks, the director of the Center for Alternative Mining Policy Development in Madison, had authored the book, *Land Grab: The Corporate Theft of Wisconsin's Mineral Resources*.

"Gedicks has done a ton of research on mining and the corporations," observed the Milwaukee Journal. "The mining interests do not like him at all. He is not fond of them, either." Indeed, he designed a T-shirt with a big corporate logo and printing that read, "We don't care. We don't have to care. We're EXXON. At Crandon, we're part of the problem."

If the state approved the Exxon project, it would transform northern Wisconsin overnight into one of the major mining districts in the country, Gedicks emphasized. "What happens to our tourist industry? Our clean water? Our clean air? There will be no opportunity for the public - let alone the state or federal government - to evaluate what the consequences will be." Corporate greed kept Exxon from creating a mine plan that would totally protect the environment, he charged. A better plan was possible with state-of-the-art technology. Instead, Exxon planned to rape the north woods.

"Depending upon his audience he would wear various garbs," contended Crandon mine executive, Barry Hansen. With his long hair, head band, and moccasins, Hansen thought, "[Gedicks] liked to play to the Native American community." "He practiced pseudo science," Hansen argued, and accused the Crandon mine of mining "uranium" that was "radioactive." He showed up at some meetings wearing a Che Guevera shirt and a beret, producing a "revolutionary sheik appearance," said Hansen. (Gedicks denied that he ever wore a Che Guevera shirt or dressed in Indian garb.) chic

"At public meetings he would put a tape recorder in front of me, [and] take pictures of me every time I spoke to somebody," Jim Derouin complained. He "tried to instill fear into the hearts of people." "Perception is a key thing when you are dealing with political events," J. Wiley Bragg observed of Gedicks. "Gedicks spread the illusion that there were a lot of people opposed to the project."[11]

Gedicks grew up in a staunch Catholic household in Pennsylvania. In his family social justice was an important value. For high school he enrolled at a Maryknoll Catholic seminary intending to become a missionary. Five years later he left the seminary, matriculated at the University of Wisconsin-Madison and entered graduate school in sociology and earned a Ph.D. degree in 1979. In Madison, he had his first experience with the "violent, repressive" activities stifling dissent against the Vietnam War. In 1970, as an anti-war activist in Madison, he was arrested for attempted arson of the ROTC building and spent ninety days in jail. He was convicted, but claimed to be not guilty.

Married and divorced, with no children and no teaching position, by 1984 he was impoverished, surviving on a $5,000 grant for the whole year and living in a small office in a Methodist church in Madison. The following year he secured a teaching position in sociology at the University of Wisconsin-La Crosse.

His interest in mining grew from a summer spent in Peru and his study of South American mining companies. "They had no regard for the people involved in the operations," he said.[12]

In 1982 he founded the Wisconsin Resource Protection Council, the first statewide organization to oppose sulfide mining, became its executive director, and wrote scores of WRPC newsletters. Gedicks recruited supporters in communities threatened by mining companies. The initial sixty members grew to three hundred within a year. The fastest growing local chapter was in the Town of Nashville in Forest County.

In addition to teaching at UW-La Crosse, he spent many weekends on the road recruiting and organizing and speaking, and did the same in the summer months and during sabbaticals. Over Christmas vacations and spring breaks, he wrote dozens of anti-mining commentaries for Wisconsin newspapers. "I essentially gave up my personal life," he said.

Called "Dr. Al" by his supporters, Gedicks was small in stature and appeared unassuming. Within the anti-mining movement, however, he gained stature and was admired for his research on mining companies, for his articulate voice and pen, and for his organizing skills and determination. "You manage to keep [Exxon] off balance!" praised Evelyn Churchill.[13]

Wisconsin's public interveners became concerned about the lack of public communication regarding the mine. Kathleen Falk became the public intervener for Crandon in May 1983 after seven years as co-director and counsel for Wisconsin's Environmental Decade. She and her successor, Waltraud (Wally) Arts, complained to the DNR - in unusually harsh language - that the public was not being adequately informed about the mine proposal. Only Exxon and the DNR knew what was happening. "The result of your staff's position is to prevent the public from being able to have any effective say in the daily ongoing decisions made on the applications," Falk complained to Carroll Besadny, the DNR secretary. "Right now, the decision making process is entirely lopsided: DNR and Exxon - That's all whose voices count."

The public intervener should be given "party status" immediately and not have to wait until just before the Master Hearing, Falk argued in March 1984. "Only Exxon and your Department will be prepared for the hearing, having had years of discovery and give-and-take. That is blatantly unfair and does not further any objective of the adversary system or the mining laws."[14]

In 1985 Wisconsin's Public Intervener Advisory Committee held a hearing in Crandon to find out how local residents viewed the mining project. Chairman Ron Koshoshek concluded that the public was "getting edgy" and was "desperate for information" on the permitting process. Many of the eighty people who attended the hearing complained that the state had been studying and restudying the same questions for two years with no new answers.

Wally Arts, the new public intervener, contended that there were still unresolved questions about the project. What were the effects of pumping to keep water out of the mine? What kinds of monitoring and safeguards would be used for the mine's tailings ponds? Groundwater presented unique problems, Arts contended. "It's not easy to fix groundwater. Once an aquifer becomes contaminated, the ways technology can deal with that are quite limited and extremely expensive." "The company has shown little or no effort at serious contingency planning for mitigation measures," Arts complained in April 1985. "In light of the company's constant statements that the permitting process is too slow, it is difficult to understand their foot-dragging on this issue." Arts wanted the Department of Natural Resources to take "a long, hard look" at Exxon's proposal for the tailings ponds. The technology was untried, and the tailings ponds would be the largest above-ground landfill in the state.

But Arts didn't push Exxon or the DNR too hard. Since two previous public interveners, Peshek and Falk, had "blessed" the mining laws and rules, Arts had no intention of blocking Exxon's permits. "It was a matter of mine on what terms," she recalled. "Not mine or no mine."[15]

*   *   *

The Sinsinawa Dominicans, a congregation of nuns headquartered in the southwestern corner of Wisconsin, were affiliated with the Interfaith Center on Corporate Responsibility (ICCR), a New York based coalition of 250 Protestant and Catholic groups that committed their investment dollars only to socially accountable corporations. The Dominicans had carefully purchased two hundred shares of Exxon stock, the minimum amount allowing them to prepare a stockholder resolution and deliver it at Exxon's annual stockholders meeting.

Sister Toni Harris and her Dominican colleagues were inspired by their congregation's founder, Samuel Mazzuchelli, an Italian-American missionary. In the first half of the nineteenth century, Father Mazzuchelli helped build twenty churches in the tri-state area of Illinois, Iowa, and Wisconsin, and also founded parishes, schools, a college for men, a new Dominican province, and in 1847, a religious congregation of nuns, the Sinsinawa Dominicans.

He affirmed the rights of Native Americans, conducted prayer services in native languages, and wrote letters to government officials denouncing the treatment of natives and the failure of the government to respect treaty rights. "Given the legacy [of Mazzuchelli] we thought it was perfectly appropriate for us to continue to ask questions about what was happening to the native people," observed Sister Harris. Furthermore, the order was committed to work for justice. "Working for justice was part of our mission statement," said Harris. The group selected the Mole Lake Chippewa as the tribe they wanted to assist.[16] They sent representatives to Exxon's annual stockholders meeting in Orlando, Florida in May 1983. "We see our shareholder actions as a vehicle to give access to corporate board rooms for communities like Mole Lake," Harris explained.

"This should add some uncertainty for Exxon about the viability of the project," Al Gedicks wrote George Rock before the Orlando meeting. (Gedicks was assisting the nuns.) "I doubt if they will be expecting anyone to raise the issue from the floor of the meeting. It should be fun."

"For the most part, annual meetings are settings in which corporate officials report past financial performance and make vague predictions about the future," reported the Orlando Sentinel. Exxon spent $200,000 and brought in seventy-five employees to staff the meeting, attended by 1,800 people. The Sentinel claimed the nuns "livened" up the normally "dull" gathering.

To prepare for the annual meeting, Arlyn Ackley and four others from Mole Lake drove through the night to Orlando and met with Sister Harris at a Denny's restaurant to discuss their plans. The resolution they drafted simply requested a delay in the Crandon mining project pending settlement of Sokaogon Indian treaty claims. Harris and Ackley each spoke for three minutes at the shareholders meeting, and Harris thought Exxon was "surprised" and "embarrassed" by their appearance. Her resolution received 2.5% of the shareholder's vote.[17]

Fred Ackley

The following year the nuns returned with a resolution for greater Exxon investment in improved pollution control technology to protect the Sokaogon's wild rice, but Exxon's attorneys claimed the Securities and Exchange Commission should disqualify the resolution by using its proposed rule revision that allowed the SEC to dismiss a stockholder resolution if it affected less than 5% of the company's revenues or sales. "This is a major setback for the Sokaogon," said Sister Harris. "It closes off yet another avenue in this tribe's efforts to have their voice heard in the decision-making process."

Consequently Gedicks composed a press release about the resolution and sent it to national news networks. "Exxon may have prevented us from addressing their stockholders at their annual meeting, but we may do them one better if we can address their stockholders through their living room TVs," Gedicks told George Rock.

Sister Harris felt Exxon's top thirty shareholders should be contacted regarding the Sokaogon. "As you can see, most shareholders are other corporations. However, Sr. Mary Ellen [our treasurer] indicated that some of those organizations may be sympathetic. (e.g., teachers, retirement organizations, universities, etc.)"[18]

On February 23, 1984 several Exxon vice presidents met with the nuns and a delegation of supporters at a hotel near O'Hare Airport in Chicago. Ruling on the SEC's 1984 shareholder resolution was still pending. "Apparently, this meeting with us is being held just in case the SEC does not uphold Exxon's appeal. (They're covering both bases.)," wrote Harris before the meeting. "They also want to meet with us before the printing deadline for their annual report; if we agree to withdraw, then they won't have to print the resolution in the pamphlet and be 'embarrassed' in front of all their shareholders!" Then Harris gave advice to members of the delegation. "Let's not be taken-off guard by big ol' Exxon's request to have a meeting with little ol' us. I'd encourage us to keep the meeting fairly formal. (E.g., try not to let Exxon representatives move into a first-name, 'folksy' basis. Such a manner can be misleading.) Let's be polite, but definitely cautious."[19]

Before Exxon's annual meeting in Los Angeles on May 16, 1985, the nuns issued a press release quoting Sister Sarah Naughton, the current treasurer of the Dominicans. She stated Exxon had published "some beautiful public relations type booklets - expensive, glossy paper full-color photos embossed covers." Harris added that "rather than spending money on those kinds of public relations efforts, we'd like to see a simple, concise report with clear direct answers to our questions about the impact of mining on native peoples."

The nuns remained undaunted. A 1985 resolution was more powerful and all-encompassing than the previous ones. It included the following:

What is our Company's policy regarding claims by indigenous groups to lands on or near which our Company has a mining operation?

In view of the potential environmental risks of mining operations, what efforts is our Company making to minimize these in the localities of its operations, in plans for reclamation and for pollution abatement?

In view of the international nature of our Company's mining activities, does our Company use a consistent set of environmental protection standards both within and outside the United States?

For each current mining operation, describe our Company's relationship with the governments, with indigenous groups, and with private citizens in the mining area.

Describe the nature of and reason(s) for any public opposition to our Company's mining operations wherever this may occur.

The resolution received 3.1% of shareholder votes. After a brief hiatus, the nuns would continue their efforts to educate shareholders and embarrass Exxon.[20]

Mine opponents enjoyed only modest success up to 1986. COACT Research of Madison polled residents in the town of Lincoln and found substantial concern about "potential lake and groundwater pollution from the mine, along with potential for boom town conditions, increased property taxes and noise."

At the 1983 annual meeting of the Town of Nashville, 41% of voters favored a moratorium resolution. Altogether about eighty small Wisconsin communities passed moratoriums against metallic mineral mining. So did many of Wisconsin's Indian tribes, but in 1986 opposition to the mine was quite ineffective.[21]

There was very little anti-mining pressure, reflected DNR's Robert Ramharter. The opposition was poorly organized or funded, he thought. "Ninety-five percent of the pressure was to move the project ahead." The three nearby tribes were not heavily involved nor was the Sierra Club. "We were not well organized," agreed George Rock.

Opponents had reacted to each step of the process instead of using concerted offensive strategies. When Exxon or the DNR released a report, opponents responded. "There was very little pro-active opportunity," said Tomahawk anti-mine leader, Jim Wise. Opponents sent out press releases, held meetings, and attended hearings. "No one was carrying a sign," said Wise; "No one was raising hell."[22]

* * *

In May 1986, the DNR issued a draft of its Environmental Impact Statement and scheduled a public hearing at the Nashville Town Hall in June of that year. Three hundred and fifty people attended, and eighty testified.

Erhard Huettl, chairman of the Forest County Board, claimed most area residents favored the mine. He cited hundreds of new jobs as the most attractive feature. Of course, residents didn't want their groundwater damaged, Huettl said. "But they, like I, feel secure that if this is monitored properly, it will be safe."

To the nearby Indian tribes, the mining project was very disturbing. "The water of Swamp Creek gives our reservation its life - wild rice," said Francis Van Zile, a Sokaogon Chippewa of Mole Lake. "You're murdering me, you're killing [Mother Earth]. It hurts me to know you want to come in and destroy us."[23]

In mid-November 1986, the DNR issued its Final Environmental Impact Statement (FEIS) and endorsed the project. The primary concern in the 446-page document was whether Exxon had effective plans to preserve water levels of lakes and streams near the ore deposit. The DNR wanted Exxon to do more to solve the problems caused by pumping water from the mine.

Without mitigation, mine pumping would significantly alter the levels of Little Sand Lake, Deep Hole and Skunk Lakes. The most serious damage would affect Little Sand Lake which would experience an average lake level decline of approximately 6.9 feet.

"Exxon proposes to mitigate surface water impacts by pumping water from a system of interception wells near the mine in the glacial aquifer, from the uncontaminated groundwater collection system within the mine and from dedicated wells near streams," the FEIS reported. But Exxon's plan might not provide sufficient water to mitigate all impacts to surface waters. "Pumping groundwater from around the mine and from dedicated wells near streams for mitigation would further reduce groundwater contributions to surface waters, potentially resulting in a shortfall of mitigation water." Clearly the DNR did not regard the problem as insurmountable. Linda Bochert commented that the issue of pumping water from the mine was "solvable."

On many of the potentially controversial issues, the DNR's FEIS found little fault with the mining company's plans. The report did not anticipate adverse effects on the tourist industry or outdoor recreation associated with the Wolf River. Wastewater discharged to Swamp Creek would contain sulfate and "very low levels of various metals," but no long-term impacts to the aquatic life of the Wolf River were "expected." This included "no significant changes in heavy metals, no change in Wolf River water temperature, and no reductions in river flow in the reservation."

The company had designed project facilities to contain spills and leaks. In any event, "spills with resultant environmental impacts would be a low probability event."

Seepage of contaminants from the tailings ponds was a "low probability low risk event." Monitoring to determine seepage would occur throughout the operation. "If increased seepage should occur or appear imminent, the evaluation and assessment procedures detailed in the contingency plan would be implemented."

Nor would mining adversely impact Rice Lake. "Surface construction activities, located two and one half miles upstream from one of Rice Lake's inlets (Swamp Creek), would have erosion and runoff controls," the report said.

The quality of water following the mine's closure should be similar to pre-mining conditions. "Contamination of the glacial aquifer is not expected to present a risk." If contamination did result, the groundwater contingency plan would be followed. Periodic inspections should limit leakage.[24]

The FEIS delighted supporters of the mine. "I'm glad to see it. We need the employment," said Crandon Mayor Kenneth Aubol. "I think the only thing we can do is take the word of the DNR. They're supposed to be our professionals. If you can't trust them, I don't know what we're going to do."

The FEIS candidly outlined some potential consequences of the mine that could negatively impact the Mole Lake Sokaogon, but in endorsing the mine proposal as a whole, it ignored those tragic consequences. Mole Lake and the Potawatomi Indians who lived near the mine would experience "some adverse socio-cultural impacts and limited employment opportunities," the report said. The Sokaogon "lack relevant work experience, education, and skills" necessary to work at the construction site, and reluctance to work underground "would be another constraint." A survey of Mole Lake residents showed that a majority had no desire to work at the mine.

The influx of many non-Indian workers and their families into the area could result in the Sokaogon being a bigger minority. "Racism could become more pronounced rather than diluted." Traditional customs of the Sokaogon might be threatened. "Increased crowding on the land and more intense utilization of resources by newcomers in the vicinity of the reservation could degrade the reservation environment."

Tribal members could experience psychological stress. "This distress centers on their perception of possible negative consequences to Rice Lake and the impact the mine could have on traditional life styles and customs." Questionnaires indicated a "feeling of helplessness in dealing with potential mine development, worry about increased traffic on State Highway 55, and concern for the long-term quality of drinking water."

At the onset of construction of the mine, distant tribal members may return to the reservation in search of jobs, creating a housing shortage. More families would need to share homes. "An additional burden would be placed on waste removal, health and human services, tribal administration, and related services."

"Issues of traditional versus new values, money economy versus subsistence, taking from the land or being part of it would involve personal decisions as well as tribal concerns. Issues related to the project could become the chief divisive issue among the tribal members. The concerns would be difficult to quantify or remedy, but would be carried with the tribe even after the mine had closed."[25]

Dave Blouin   Al Gedicks

Al Gedicks castigated the DNR for its narrow view of the importance of wild rice to the Sokaogon. His report stressed the economic significance of wild rice in providing income and employment to tribal members. "Wild rice also has cultural and religious significance," Gedicks noted. "The Chippewa consider wild rice to be a sacred gift. Wild rice is always served at important community activities and is a vital part of the celebration held with the naming of a new child."

The FEIS often used tentative or vague terminology - "probably," "could be," "not completed," "projected," "effects will be minimal," "impacts are uncertain," and "effects are unknown" - that did not instill confidence in the DNR's conclusions.[26] "Reading the document leaves the impression that there will be no adverse environmental consequences. That's nonsense!" said Wally Arts.

The Environmental Protection Agency found major shortcomings in the FEIS. "There is a consistent pattern of failure to deal with the long-term potential impact of this project on the biota and natural resources of the area." The project lacked adequate contingency plans to assure environmental protection. "Some form of performance guarantee from the Exxon Corporation may be appropriate and should be considered." Nor had the DNR addressed the effect of long-term discharge of heavy metals to Swamp Creek and ultimately to the Wolf River.

The DNR considered the failure of the tailings ponds a low probability. But if there was a failure, said the EPA, did Exxon have a plan for dealing with it? "Does this plan contain preventative measures and steps to be taken to curb contamination of the ground and surface waters affecting the water quality of Hemlock Creek, Swamp Creek, and, eventually, Rice Lake?"

The EPA worried that the mine wastes discharged into Swamp Creek would have long-term cumulative effects on trout in the Wolf River. "Lead and mercury in the effluent alone are more than ten fold higher than USEPA water quality criteria....The DNR only considered the effects on aquatic life and fish when subjected to one toxin at a time when, in fact, they will get all of the many toxins at once. No concern was given to the accumulative effects or acute worst case possibilities."[27]

The next step in the permit processing would be the "Master Hearing," a court-style procedure where witnesses and lawyers debate the issues of building the mine. But that phase never occurred because Exxon made a momentous decision.

*　　*　　*

Barry Hansen's demeanor had changed, thought Edie Franson, a secretary in the mining company office in Rhinelander. He was quiet and thoughtful. Soon she learned why. "A decision has been made by Exxon Minerals to shut down [the mine operation]," Hansen informed the staff at a meeting. The company's announcement on December 10, 1986, stressed the reason for the pullout was the lackluster price of minerals. Zinc was selling for 44 cents and copper for 63 cents a pound. Zinc had sold for 80 cents in 1974 and 55 cents in 1980. Copper had peaked at 96 cents a pound in 1980.

"I am very upset Exxon is not going to stay in Wisconsin," Governor-Elect Tommy Thompson said. "You can't measure it in dollars and cents. There were so many people who had their hopes up."

Mayor Ken Aubol had his hopes dashed. "Although we never had the mine, it puts us back to square one," he told a reporter. "It's a cold day in more than one way," he remarked as he cast a brief look at a storefront thermometer that recorded a twelve degree temperature. "I think that people here were beginning to believe that the mine would come true and that would have meant a great deal to the local economy."[28]

Although the company cited low metal prices as the reason for the decision, pro-mining forces found another. State Senator Lloyd Kincaid blamed the DNR for taking too long to review the project. Too much bureaucratic red tape caused Exxon to pull out. Another factor, Kincaid said, was the opposition of "hostile" environmentalists who wanted to "close" northern Wisconsin.

Erhard Huettl echoed Senator Kincaid's remarks. "A lot of people think that mining regulations and the complaints of environmental groups finally drove Exxon away from Wisconsin," Huettl said.

Columnist Dick Timmons of the Rhinelander Daily News hoped Wisconsin would be more receptive to welcoming new business. "Let our regulating bodies move with more dispatch in reaching their decisions and not make a career out of the process. Let the Indians allow their more mature leaders to take an active part in working with the mine operator in order to bring the highest benefits to the tribes and families. And let true environmentalists speak for the concerns of our lands and resources, rather than professional special interest social advocates who, if they had their way, would have prevented the invention of the wheel."[29]

Many years later Barry Hansen cited another reason for Exxon's pullout. In addition to the price of metals, top officials at Exxon lacked a commitment to the Crandon project. "We were six weeks away from the final hearing," said Hansen. "We had a FEIS....and had preliminary approvals from the towns of Lincoln and Nashville. A mine with a permit is much more marketable than a mine without a permit."[30]

While pro-mining forces bemoaned the decision, opponents celebrated. "It's already a great Christmas," beamed Sonny Wreczycki. Arlyn Ackley, tribal chairman for the Mole Lake Chippewa, called Exxon's announcement a victory for the tribe. "You might say that this has been a thorn in our side for more than ten years." "This will be the first Christmas season we can truly celebrate in a long time," Al Gedicks wrote George Rock.[31]

Few people noticed the company was not totally abandoning the project and revealed no plans to pull out of Wisconsin entirely. Exxon will "maintain a presence" in the state, Hansen said.

Hansen's coda prompted a warning from DNR Secretary Carroll Besadny. He told David Hase, an attorney for Exxon, that if the corporation decided to continue the project sometime in the future, it should not expect the DNR to get the permit process restarted immediately. "We'd have to re-tread and get the project going," Besadny said. "We can't just turn things on and off...I hope everybody will remember this meeting today, that we told you so."[32]

The question loomed. Why didn't Exxon proceed with the permit process? Exxon had taken the preliminary steps in the process, and if the company wished to revive the project in the future, why not get the permit and just hold on to it until circumstances improved?

"In hindsight, that would have been the smartest thing they could do," said Gordon P. Connor. "Then they could attract a partner or sell it. They'd have the permit, and it would have been worth a lot more money and would have ended the circus that followed."[33]

Many Exxon employees in Wisconsin agreed. "The consensus among many of Exxon's attorneys and all of Exxon's outside advisors was that this was a terrible mistake," recalled an Exxon consultant. Attorney David Beckwith of Foley and Lardner,

Exxon's primary Wisconsin law firm, tried unsuccessfully to convince Exxon to proceed with the process, get the permits, and then deliberate on the future of mining in Crandon. Exxon's decision bewildered many people. "It strikes me as almost bizarre to get to this point and then give up," said Attorney Kevin Lyons. "That was a bad decision," said Robert Ramharter. "They could have secured the permits....I don't understand why they did it."[34]

CHAPTER FOUR

INTERIM 1987-1993

After Exxon abruptly halted the Crandon project in 1986, George Rock and others critics boxed up their mining literature. Opposition to the mine had been all consuming for Rock. "There was a reprieve for us to go on with our [normal] lives."

The Crandon mine went into abeyance - no engineers on site, no permitting activity, no controversy. The media diverted its "mining attention" elsewhere, particularly to the Ladysmith area where Kennecott had resumed its attempt to mine.

Back in 1976 local opponents, led by the Churchills, had stymied Kennecott by denying the company local zoning approval for its mine. A decade later, after deciding to reapply, Kennecott and British corporate giant, Rio Tinto Zinc, the parent companies of Flambeau Mining Company, used a different approach. Flambeau Mining scaled down the project and developed a sophisticated strategy to neutralize the Rusk County anti-mining ordinance.[1]

In 1987-1988 Kennecott lobbied intensely and successfully for a new Wisconsin law allowing mining companies to negotiate "local agreements" with local governmental authorities holding zoning authority. Now the company could identify local concerns and negotiate cooperatively with officials to address those concerns in a local agreement. (Such concerns could be hours of operation, monetary payments, hiring, water well compensation, property value guarantees, and noise reduction). The agreement had to be approved by the local municipality in a public meeting following a public hearing on the matter. In the case of the Flambeau project, three local municipalities would have to approve the agreement and could receive financial benefits.

This "local agreement" law assured Kennecott they had local zoning approval <u>early</u> in the permitting process. If Exxon ever decided to resume its Crandon project, they could use the same approach.[2]

Flambeau Mining gave the DNR its notice of intent to collect data in July 1987. Three-and-a-half-years later the master hearing examiner granted permits to the company to mine the deposit.

Critics of the Flambeau mine envisioned "Appalachia, strip mining, slag heaps and befouled rivers." But local supporters of the mine wore buttons saying "Rusk County Yes, Protesters No." The mine won the endorsement of the DNR and respected consultants. Bob Ramharter, the DNR's Flambeau project manager, said the environmental impacts were relatively minor, and he had "no doubt that it's a safe project."

The public intervener hired a hydro-geologist, Douglas Cherkauer, a Professor at the University of Wisconsin - Milwaukee, as an independent expert to review the Flambeau mine. Cherkauer, who had earlier criticized the Crandon project, endorsed the mine in Ladysmith. His judgment was that groundwater contamination would occur adjacent to the site, "but that it will be confined to a very small area which will not directly affect humans and which will produce no measurable changes in the quality of the Flambeau River."[3]

The open pit mine was 50-feet wide, 2,600-feet long and 220-feet deep. Opposition to the mine focused on its sensitive location. The DNR granted the company permission to mine within 140 feet of the Flambeau River, rather than the 300 feet from a navigable river normally required. Critics predicted that wastes would contaminate the river, but the water treatment process the company used turned out to be very successful. The Wisconsin standard for copper levels in water was 42 parts per billion; the treated water from the mine had a copper level of only 5.82 parts per billion.

"Generally, dealing with Flambeau was very good," said the DNR's Larry Lynch who studied the project. "Flambeau Mining was very responsive to our concerns," he said. "Their wastewater people were proactive and always tinkering to make it better."

The company hired about eighty percent of its thirty-three employees from the local area. Ladysmith was economically stronger than it had been in several decades, its mayor contended. John Terrill, managing editor of the Ladysmith News, lauded the local agreements. The pacts had two hundred special conditions the mining company had to meet, "and the company lived up to those conditions," Terrill said. Opponents "used any excuse to oppose the mine," he added. "They made appeals that were misrepresentations. They used scare tactics."[4]

Some of the criticism of the Ladysmith mine from people opposed to the Crandon mine was accurate. Roscoe Churchill pointed out that Flambeau's owners reaped $500 million in profits, but paid only $27 million in taxes. "That is a rip-off," he said. Opponents could also take some credit for the extensive safeguards built into the project.

However in the main, the critics were confounded. Heavy metals such as mercury and arsenic did not leach into the ground and surface water. No boom-and-bust cycle caused by the mine's presence upset the local economy. Instead of driving away tourists, the mine actually attracted thousands of them. "The Flambeau Mine is an eco-extremist's nightmare," said a mine supporter, "mainly because they have absolutely nothing to complain about."

Opponents usually blustered, unwilling to concede that the mine could operate safely. Roscoe Churchill remained fixated on Kennecott. "Kennecott must be stopped, they are such buzzards," he wrote George Rock.[5] But Churchill and his allies were left with only dire predictions no one could prove. "I won't live long enough to see [mine pit water] get into the landscape, but it will," he lamented.

Flambeau's mining site closed in June of 1997, and reclamation began immediately. The Flambeau mine produced 181,000 tons of copper, 334,000 ounces of gold, and 3.3 million ounces of silver.

Actually the Ladysmith mine held few analogous lessons for the Crandon project. The Crandon mine would be larger and more complex than the mine in Ladysmith. Unlike Crandon, the ore taken from the Ladysmith mine was shipped to Canada by rail for processing. The Flambeau mine was above ground and primarily a dry operation; Crandon would be an underground mine built near lakes and wetlands, and the mine shaft would need to be pumped clear of water, potentially lowering area lakes and drying up wells.

Nonetheless, the approval and success of the Ladysmith mine encouraged Exxon. Ladysmith set a precedent, and the new local zoning rule would help them immensely in negotiating their own permitting process.[6]

*   *   *

Several other events affected the future of the Crandon project. Before 1989, Wisconsin had no provisions to protect high-quality waters as required by federal regulations, leading the Wisconsin Natural Resources Board to direct DNR staff to develop new anti-degradation regulations. In 1988 the Board reclassified state rivers and granted the upper Wolf River its highest classification as an "Outstanding Resource Waters," meaning that by law the Wolf River must stay as clean as it was currently. Any treated water dumped into the river must be as clean as the river water. The protection extended to the Wolf's tributaries, as well, most notably to Swamp Creek at Mole Lake.[7]

In a letter May 20, 1988 to DNR Secretary Carroll Besadny, from Exxon's environmental and regulatory affairs manager, James Patton, strongly protested the Wolf's classification. The action "could be interpreted to prohibit any changes to the water chemistry of one portion of the Wolf irrespective of whether such change would be barely detectable [if at all] or would pose any threat to aquatic life or other uses of the river." Hinting at Exxon's future intentions, Patton concluded that "It could create a significant potential roadblock to any future resumption of the Crandon project" and would damage economic development in the area.

An editorial in the <u>Milwaukee Sentinel</u> disagreed with Patton's contentions. "Company officials are afraid that protecting Swamp Creek at that level could damage or bar their plans for mining....If Exxon officials are serious about the mine and about keeping the environment clean, as they have said in the past they are, then they should find a way to build the mine without damaging Swamp Creek."[8]

*   *   *

Shortly after midnight on March 24, 1989, an oil tanker, the Exxon Valdez loaded with North Slope crude oil, ran aground on Bligh Reef in the northeastern portion of scenic Prince William

Sound, spilling about eleven million gallons of oil into the sea. It was a non-mining disaster that sullied Exxon's overall reputation for respecting the environment.

The spill devastated the delicate food chain that supported Prince William Sound's commercial fishing industry. "Among the many casualties," said one report, "were 2,800 sea otters, 300 harbor seals, 250 bald eagles, as many as 22 killer whales, and an estimated quarter-million seabirds." Eventually the oil "contaminated a national forest, four wildlife refuges, three national parks, five state parks, four critical habitat areas and a state game sanctuary."[9]

The oil industry had previously contended that technological advances made it possible to contain oil spilled in offshore accidents, an argument used to justify new oil exploration and development in environmentally sensitive areas, like Alaska.

In 1969, Prince William Sound's Chugach natives had not insisted on money to use the port of Valdez. Instead, they demanded only commitments from the oil companies to use the most advanced anti-pollution methods to protect fisheries, wildlife and migratory birds. Exxon and other oil companies had promised they would use state-of-the-art radar, place oil-spill equipment throughout Prince William Sound, and maintain emergency response crews trained for emergencies.

But in March 1989, when the Exxon Valdez spilled its oil, the oil company's emergency equipment didn't arrive on the scene for thirty-six hours. Disaster investigators later discovered oil companies had misled government inspectors by creating spill response teams made up of pipeline workers who had no training or experience in dealing with an oil spill.

The Exxon Valdez had sophisticated radar on board, which should have made it impossible for the tanker to drift onto Bligh Reef, but the radar wasn't turned on, the crew didn't know how to use it, and it was out of order anyway.[10]

An enraged Alaska fisherman blamed Exxon for not cleaning up the spill quickly enough. "The bottom line is, we're screwed," he said. "There's no way you can win, dealing with Exxon." At a press conference, protesters carried signs reading "Exxon Double Cross" and "Don't Believe What You Hear."

A Coast Guard study revealed eighty-five percent of all tanker accidents were the result of human error. Could human error plague the Crandon mine? critics in Wisconsin wondered.[11]

In 1994 a federal jury in Anchorage ordered Exxon to pay $5 billion in punitive damages to 10,000 fishermen and other Alaskans for ruining 1,200 miles of Alaska's shoreline.

Subsequent events cast doubt on the trustworthiness of Exxon. After the court decision mandating Exxon to pay the $5 billion, it came to light that the company had concocted a secret scheme to avoid paying all the damages. In 1996, disclosure of secret agreements in the oil-spill case that would have allowed Exxon to reduce its punitive damages by hundreds of millions of dollars shocked legal ethicists. The Wall Street Journal reported that U.S. District Judge H. Russel Holland of Anchorage, Alaska, discovered that Exxon cut secret deals with seven Seattle-based fish processors to refund to Exxon about $730 million of the $745 million of punitive damages they claimed in the case. "The arrangement infuriated Judge Holland, who called it an 'astonishing ruse' to 'mislead' the jury in the case and 'negate' [the] 1994 verdict." The agreement would have allowed Exxon to reduce its $5 billion in punitive damages by 15%.

"Exxon intended to share in the very punitive damages award that the jury deemed necessary to fulfill society's goal of punishment and deterrence," the judge said. Exxon had "acted as a Jekyll and Hyde" in the matter, "behaving laudably in public, and deplorably in private." Legal ethicists had never encountered such an agreement. One called it "an apparent fraud on the jury."

Exxon clearly understood the battering its corporate image suffered as a result of the Valdez disaster, but it failed to realize the event could energize opponents of its metallic mine far away in Crandon. Many there believed Exxon unworthy of public trust.

Could Exxon be trusted to act honorably and openly on the Crandon Mine project? One expert concluded that "so long as there are ships, and humans steering them, accidents will happen, and maybe huge ones." The same could be said for the Crandon mine.

So long as there are mines, and humans running them, accidents will happen, and maybe huge ones.[12]

*　　*　　*

Prior to Exxon's mining efforts in the state, a spear fishing controversy arose in northern Wisconsin. Anglers demeaned and defamed Indians who asserted their tribal right to spearfish. That those opponents would become allies and view the Crandon Mine as a danger, no one, including Exxon, anticipated.

In the nineteenth century, when the United States negotiated treaties with the Indians, the tribes agreed to give up much of their land in exchange for assurances that they could continue to fish at their customary places within the lands they had agreed to cede.

Subsequently, the courts ruled that these rights still existed, that tribes were "entitled to substantial shares of the fish that can be harvested, and that states generally cannot regulate tribal members in the exercise of these rights."

A series of court decisions in the early 1980s confirmed the right of Wisconsin Chippewa (Ojibwa) to hunt and fish on ceded territories. Thereafter, spear fishing occurred in a number of locations in northern Wisconsin, but mostly on lakes near the Lac du Flambeau Reservation.

Some groups in Wisconsin, primarily Protect Americans' Rights and Resources (PARR) and Stop Treaty Abuse (STA), opposed the Chippewa's spear fishing. In his book, <u>Chippewa Treaty Rights,</u> author Ronald Satz noted that opponents thought it was "unjust" for the Chippewa to have special hunting and fishing privileges denied other Wisconsin residents just because of some "old treaties." Indians had "more rights" than white citizens, irate critics argued; Indians and whites should have "equal" rights, and treaty rights should be abolished. Some non-Indians feared that Chippewa's spear fishing would exhaust the fish population and leave little for tourists, even though the band speared only two percent of the annual 670,000 walleye population.[13]

Violence erupted in 1985 when opponents threw rocks at Indians spear fishing on Butternut Lake in Ashland County.

Violence, verbal abuse, and threats peaked in the spring of 1989. The protest activities "included death threats, use of effigies, rock throwing, use of wrist rockets (a lethal, sling-shot type weapon), pipe bombs and harassment of fishermen on the lakes by creating large wakes," said one report. "Some tribal fishing boats were swamped. Ojibwa fishermen were hit with rocks or other objects both on and off the water. Treaty supporters and Ojibwa fishermen received death threats." "They threw rocks at me," said Mole Lake's Fred Ackley, "and spat on my wife and children, calling them names. I hurt from that. [I suffered] mental anguish."

Racism was palpable. Verbal taunts included "Save a walleye, spear a squaw;" "Spear a pregnant squaw, save two walleyes," "Custer had the right idea," "Scalp 'em," "You're a conquered nation, go home to the reservation," and "diarrhea face." A bar owner posted a sign, "Help Wanted: Small Indians for mud flaps. Must be willing to travel." Riot squads from metropolitan areas had to assure safety for Chippewa spear fishermen and their families.[14]

The Chippewa attracted positive support from other state tribes and from non-Indian civil rights and conservation groups. Treaty support organizations began to counter protesters at boat landings by serving as "witnesses" of the violence and racism. "They, along with other organizations and individuals, provided peaceful witnessing at the Wisconsin spear fishing landings and offered a pro-treaty perspective to the media."

By 1991 most of the protests at the boat landings had ended, and shortly thereafter former opponents shared a common concern - preserving the environment, particularly from mining in Wisconsin. A leader in the reconciliation of anglers and Indians was Walt Bresette, a Red Cliff Chippewa and co-founder of the Witness for Nonviolence and the Midwest Treaty Network. An eloquent speaker and writer, Bresette was one of the few during the spear fishing controversy who understood that issue's environmental underpinnings. His friend, Zoltan Grossman, observed that "Even as the Chippewa spearers and the white anglers were in conflict over treaty rights, Walt understood that they really had more in common than they had differences. He felt the people of northern Wisconsin had to defend their clean environment and their mom-and-pop businesses together, or both would be lost to outside corporations." Unless anglers and Indians worked together to fight mining, said Bresette, "The only thing that's left is you and I fighting over some poison fish."[15]

*   *   *

In addition to the Valdez spill, two additional developments would adversely affect Exxon's success if they decided to resume the permitting process for the Crandon mine. The first was enhanced financial resources of the Indian tribes, and second, the tribes' protection provided by the EPA.

Until the 1990s, northern Wisconsin tribes had inadequate dollars to combat corporate projects and therefore were not prominent in the anti-mining movement. This changed dramatically with the legalization of casinos on Indian reservations in the state.

In the 1970s legalized gaming outside of Nevada and New Jersey was confined mainly to bingo halls on Indian reservations or to fund-raising for non-profit organizations and churches. With the passage of the federal Indian Gaming Regulatory Act of 1988, casinos were permitted on Indian reservations. A compact negotiated between a state and a tribe would regulate the casinos. Wisconsin tribes signed their first agreements authorizing casinos in 1991. In that year state casinos netted $100.2 million; the following year $302.4 million; and in 1993 $406.1 million.[16]

Although the two Mole Lake casinos were smaller than those of other tribes, through their operations and that of the Potawatomi tribe, the unemployment rate in Forest County dropped to five percent, a third of what it had been six years earlier. "For the first time ever, I've seen an employment agency (located) in Crandon," said Duane Derickson, tribal planner for the Mole Lake Sokaogon.

Proceeds from gaming at the casinos raised the living standards of the tribes and provided funding for health care, housing, elderly assistance, and educational programs. Most gamblers were oblivious to the way casinos' huge revenues could enhance the tribes' ability to hire staff, consultants and lawyers to oppose the mine. "[Casinos] gave us a means in which we could use our resources to help us protect our environment," said Gus Frank, chair of the Potawatomi.[17]

The federal Environmental Protection Agency also assisted the Indian tribes. Instead of simply dealing with states to secure mining permits, corporations now had to comply with federal rules and the EPA, an agency determined to protect Indian tribes, treating them like independent states.

For years raw sewage and wastes had been discharged into rivers and lakes, making water unfit for drinking, swimming, and boating. Air pollution enveloped whole communities. Toxic chemicals leaked hazardous materials into soil and aquifers. To address pollution problems, in 1969 Congress passed the National Environmental Policy Act (NEPA), and President Richard Nixon signed the law on January 1, 1970. President Nixon transferred fifteen units from existing organizations into the Environmental Protection Agency, the new independent super agency.

Suddenly with little warning and no consultation, states were forced to accept and implement federal directives on environmental issues. The EPA established ten regional field offices, increasing its sensitivity to local issues, and facilitating active federal enforcement.

In the EPA's view, the states' judgments about environmental quality were unduly affected by local economic interests. "States would continue to play their primary regulatory roles, but within the confines of federally-dictated programs and subject to EPA's preemptory power ensuring recognition of national environmental quality interests," said James Grijalva in his history of the EPA.

In the early 1970s, the environmental legislation passed by Congress completely neglected Indian country. The EPA judged that states generally lacked adequate regulatory authority to protect Indian tribes. "Congress' silence in the environmental laws of the early 1970s convinced EPA that an unacceptable regulatory void existed in Indian country," said one authority. "The solution, EPA posited, was an approach according tribal governments a regulatory role akin to the role states enjoyed."

Although EPA had no experience dealing with Indians and no mandate to take responsibility for tribal trust assets, it did so, becoming the first federal government agency to adopt an official Indian policy. In 1983 President Ronald Reagan announced his support for tribal self-determination. The following year, EPA also espoused its commitment to tribal self-determination.

The EPA's Indian policy declared that the EPA or tribes, not states, should implement federal environmental laws on Indian lands, and it authorized EPA to cooperate with and assist tribes in developing and implementing tribal programs under federal environmental statutes.

In 1991 EPA established a process for tribes to set their own water quality standards under the Clean Water Act, which had been amended four years earlier to allow EPA to treat tribes like states. The state of Wisconsin and the mining company would be shocked when the EPA stretched its muscles.[18]

*   *   *

Two years after Exxon suspended their attempt to get permits for the Crandon mine, the company reconsidered. In 1988 Exxon told their Wisconsin attorneys the company might soon resurrect the Crandon project, but they hesitated after the Exxon Valdez disaster. Then, in 1991, Exxon requested a comprehensive report from their attorneys on restarting the mine project. What were the key legal issues? What was Wisconsin's political climate? Exxon wanted to know.

The company also began to look for a partner to share development costs for the mine. When Wisconsin businessman and lumber baron, Gordon P. Connor, put together an investment group in 1991 to purchase the mine, Exxon rejected the bid. In May 1992, Exxon and another mining company, Phelps Dodge, entered into a preliminary partnership to develop the mine, but the following November, Phelps Dodge withdrew.[19]

But back in 1986, shortly after Exxon mothballed the mining project, another suitor, Rio Algom of Canada, quietly began to study the idea of a partnership. Rio Algom, headquartered in Toronto, had mining interests in copper, potash, and uranium, and had coal mines in North America and Chile. "Exxon has had difficulty developing the [Crandon] property for many unknown reasons," an executive at Rio Algom observed.

Officials at Rio Algom sought the advice of Barry Hansen who had recently ended his employment with Exxon. Hansen told them a "significant portion of the project delays were attributable to Exxon" because the company had modified the mine plan, delaying the DNR's regulation schedule. "Also, Exxon's internal bureaucracy was largely to blame."

Hansen said the situation now looked more hopeful. The new Secretary of the Department of Administration, James Klauser, was Exxon's former lobbyist, and Governor Tommy Thompson was "vocally pro-business in general and pro-mining in particular." Hansen commented that Rio Algom's hesitation was understandable, but that in "operation, when the community gets money in their pocket, they will be less negative."[20]

Rio Algom identified critical success factors in order of importance. "Can the Crandon project get a permit and, if so, how long will it take? By how much can we reduce Exxon's estimate of the capital costs to bring this mine into production? Can higher grade zones be mined in the early years to enhance capital pay-back?" The company concluded that Crandon was a "potential world class deposit which can be profitably developed in a manner different to that proposed by its owner, Exxon."

"At this stage of our review we believe that a three-year time frame to obtain a permit is probably realistic," wrote a Rio Algom executive in March 1993. "Thereafter, construction could easily take another three or four years. Given such a long lead time, the earliest Crandon could be in production is 1999/2000. Thus, our current perspective on Crandon is that it could provide Rio Algom with an attractive but distant development project."[21]

A partnership was emerging.

CHAPTER FIVE

THE NEW CRANDON

MINING COMPANY

In 1993, Exxon Minerals with its new partner, Rio Algom, resumed efforts to secure permits for the Crandon Mine. They were optimistic for Wisconsin's DNR had blessed their earlier venture before Exxon aborted the process and therefore could be expected to approve the new one. The Crandon Mining Company had a battery of lawyers, consultants and engineers, plenty of money, and powerful political supporters in Madison. True, opponents of the mine had been irritating, but they had not mustered much strength during the earlier permitting venture. The new company confidently predicted permits would be granted within three years.

The Crandon Mining Company remained optimistic about its progress and eventual success until problems arose when regulatory agencies imposed complex, restrictive rules. Advancing the permitting process slowed. It was frustrating when a federal regulatory agency agreed with Indian tribes in disputes with the company. Then in the midst of the regulatory wrangling, the company misjudged and failed to anticipate the negative public reaction to an engineering decision.

Company officials stood out in northern Wisconsin with their "white shirts, skinny ties and Canadian and Texan accents," a reporter observed of the new partnership.

"We're really delighted to be a part of this new effort, and we look forward to the immediate start of the permitting process," Jerry Goodrich, the president of Crandon Mining Company, announced at his first news conference in September 1993. Rio Algom and Exxon had evaluated the environmental, engineering, financial, marketing and staffing needs, and concluded that only "modest changes" were needed to the plans developed in the mid-1980s. At stake were hundreds of jobs, huge tax revenues, investments projected at three-quarters of a billion dollars, and mineral sales for Exxon and Rio Algom of $4 billion.[1]

Actually Exxon's joint venture with Rio Algom was less ambitious than Exxon's proposal in the mid-1980s. The company planned to produce 5,500 tons of minerals a day instead of the 10,000 tons projected in 1986. The mine would employ fewer workers, about 400 people full time. The company would mine 30 million tons of zinc during the first fifteen years of production, Goodrich explained. If economic and marketing considerations warranted, the company would later mine an additional 25 million tons of primarily copper, extending the mine's life to about twenty-five years.

"You better believe that I'm for mining," said Crandon Mayor Vernon Kincaid of the new venture. "I want to see jobs come to this area. I'm tired seeing the kids leave because they can't find employment." The mine supporters included Governor Tommy Thompson, James Klauser, now Secretary of the Department of Administration, and a pro-business Republican-controlled state legislature. Mine opponents assumed that Thompson and Klauser would pressure the DNR into approving the mine.[2]

Jerry Goodrich, 51, was a mechanical engineer with thirty years of experience. He had worked for Exxon Coal and Minerals Company and recently had been named Exxon's U.S. Coal Coordinator, responsible for long-range planning and project development. He was now assigned to Crandon on short notice that made it difficult for him to fully understand the Wisconsin context. It appeared his threshold for anxiety was low, but he tried to mask it.

The company's Environmental Impact Statement was similar to the one they proposed in the 1980s. The company claimed that the proposed mine would protect the natural resources of northern Wisconsin, provide permanent groundwater protection, comply with surface water and air quality standards, safeguard lake and stream levels, produce a net gain of wetlands, and enhance the local economy.

"After what is probably the most thorough environmental study ever conducted in Wisconsin, we conclude that we can build and operate a mine that provides some 400 long term jobs while preserving the environment, culture and character of the North woods," said Goodrich. CMC claimed that its environmental studies totaled more than 140,000 hours of investigation by 150 engineers, scientists and technical personnel.

The mine's tailings basins would be designed with multiple safeguards. CMC would deepen or replace, at its expense, any wells for which monitoring indicated a potential impact from the mine. The mine would have minor or no effects on local lake water levels, except for Skunk Lake, a shallow, six-acre pond on the mine property that had no fish population and no cottages.[3]

The decline in groundwater levels would be temporary. Within a few years after the company closed and reclaimed the mine site, groundwater and local lakes and streams would return to their previous levels.

The mine would result in a net gain of wetlands. Mine construction would directly affect 29.5 acres of wetlands, but these would be replaced by restoration of 57 acres of high quality wetlands on a site in a nearby county.

CMC would pay about $175 million in federal and state income taxes over the life of the mine. In the first year of operation, the mine would increase the local property tax base by about $110 million, benefiting taxpayers in Forest County, the Crandon School District, and the towns of Lincoln and Nashville. The company claimed it would spend $43 million for local goods and services during the three-year start-up period, and another $33.6 million during the life of the mine.

"Studies to date show the mine is compatible with the protection of Native American reservation resources, economies and traditional cultural properties," the report said.[4]

A major concern, however, was the groundwater flow model used by Crandon Mining Company. It was an ongoing contentious element in the minds of DNR personnel. "The model, a computerized simulation called Modflow, is meant to replicate the mine as closely as possible in order to predict the amount of water the mining and processing operation will take in and the amount it will discharge out of it." The model was exceptionally complex and attempted to predict water levels and rainfall in Forest County for the following forty years.

Initially Chris Carlson, a DNR hydro-geologist, publicly criticized the company's model, but after further discussion, the company's engineers agreed to run the model again, this time using data compiled by the DNR.

The company's chief consultant was Jerry Sevick, vice president of Foth and Van Dyke, an engineering firm in Green Bay, Wisconsin that specializes in environmental studies. As the project manager for all environmental, engineering and socioeconomic studies conducted for the project, Sevick and his assistants performed the baseline studies relating to groundwater, biology, archaeology, air quality, aesthetics, and noise. He helped prepare the project permit applications and coordinated the review of work done by over twenty-five consultants.

Don Moe, the Exxon-based technical manager for the partnership, claimed the mine would not be any harder on the groundwater system than any city or large housing development.

"The city of Antigo, roughly 45 minutes away, takes more water out of the system than the mine ever will," he said.[5]

One of Goodrich's tasks was to knock on the door of a major opponent, Herb Buettner. "My name is Jerry Goodrich. I'm the new guy here for Crandon Mining Co. Been with Exxon all my life. I know you've been opposed to the mine. I'm wondering if we can talk and maybe go fishing."

The Crandon Mining Company also sought to burnish its image by engaging in public service work. Its monthly newsletter, the Crandon Chronicle distributed to 14,000 people and organizations, often described the company's community activities. The affable Goodrich dove into community service. At Crandon's World Championship Off-Road Race, he flipped hamburgers. "When I stepped up to the burger grill, I was able to take off my 'mining company hat.' For a day, I was only a fellow with an apron and a spatula, just one of many people working together."

In 1996, the company's administrative assistant, Lynn Smith, contacted Beth Stamper at the Crandon Nursing Home with a request. Would the elderly residents help get the Crandon Chronicle ready for mailing by affixing address labels and stapling the newsletter? "As soon as Lynn came up with the thought, I knew it would be a great project for our Activities Group," Stamper said.[6]

The company also claimed to be a friend of the environment. The front cover of one issue of the newsletter pictured a sturdy man fly-fishing in a river, presumably the Wolf River. The company's Wildlife Habitat Committee proposed construction of a picnic area with Little Sand Lake access, a hiking trail with camping facilities at Oak Lake, and an outdoor classroom on Skunk Lake. The company sponsored Trees for Tomorrow located in Eagle River. Trees for Tomorrow was to provide high quality field-based natural resource education.

In one year the company donated $45,841 to 111 different organizations, most in Forest and Oneida Counties. Among the recipients were the School District of Crandon ($1500), Forest County Walleye ($300), Wabeno Lions Club ($100), and Trees for Tomorrow Scholarships ($700). The company also paid $119,099 for membership dues to twenty-two organizations.

Goodrich argued that critics misrepresented current conditions, believing them similar to those at the Rio Algom uranium mines at Elliot Lake, Ontario. "Those mines were built in the 1950s," he stated. "In the early years, acid water from the mines and tailings areas entered the Serpent River and its chain of lakes, seriously depleting fish populations. However, by the early 1970s, the problems had been identified and all discharges either stopped or were subjected to treatment."[7]

Rio Algom was currently reclaiming the Elliot Lake mines and "is committed to doing so in a way that affords permanent protection to the Serpent River and its surroundings." The company transported hundreds of local people by bus on three-day trips to Elliot Lake to learn about mining from "those who live and work in a mining community." The tour began at the Civic Center in Elliot Lake. After visiting the mining museum, tourists were directed to the mines that had operated for decades. "We leave no stone unturned on our tour," according to Lynn Smith, the tour coordinator. "We visit with local residents, eat at local restaurants, and drive through areas that have been mined and are either reclaimed or are in the process of being reclaimed." Seeing a mine firsthand "makes a big difference," said one visitor.

The company insisted on replying to every critical letter to the editor published in any newspaper in the state. "That kept [me] busy," said Dick Diotte, the company's director of public relations. In a newspaper commentary, Goodrich urged that the debate become more "productive." The issues were too important to be reduced to "rhetoric, exaggeration and pure falsehoods," suggesting that critics were guilty of all three.

The company was not asking north woods residents to choose between jobs and the environment. Both were important. There would be no mine unless the "special values of the north woods" - tourism, scenery, American Indian culture, fishing, recreation - were preserved. "That is as it should be." Wisconsin's mining laws were designed to "prevent the problems that occurred at mines built many years ago, when technology and environmental standards were nowhere near what we have today."[8]

In 1995 Goodrich contracted with a New York public relations firm, The Pires Group Inc., specialists in management of public affairs for businesses and industries. Headed by Mary Ann Pires, the firm was called in to assess the adequacy of the mining company's external relations. It studied the company's history, interviewed Crandon residents, performed content analysis of news coverage, and reviewed documents provided by the company.

"Don't pay for one more sign that says 'Crandon Mining Company' and nothing else," Mary Ann Pires told Goodrich. "It is a complete, total waste of money. I saw the one at the track on Saturday, and wanted to scream. It communicated absolutely nothing about CMC, and was combined with a bank, no less. Who do poor people hate more - mining companies or banks?"[9] One suggestion for securing pro-mining allies was to focus on the parents of children who would come of working age during the mine's projected 25-year life. "This is a prime prospect-audience for Third Party support, particularly based on our Crandon interviews," Pires pleaded. "These young people will need jobs."

"You have asked for alternative suggestions regarding CMC receiving some 'credit' for its contributions to area causes, as opposed to the recent flurry of check presentation photos," Pires wrote Goodrich on December 19, 1995. She offered several suggestions:

"If the group receiving the grant has a newsletter, ask if they'd make mention of CMC's support - in their own words.

Have photos taken, if you must, with the item or service or facility, etc., that resulted from the grant.

If the group wants to recognize CMC at its annual dinner, picnic, etc., let them!

Our problem with this recent spate of photos is how contrived and artificial and ineffective it is. The clumsy publicity can undo the good of the contribution."

The Pires Group's surveys found that 90% of people interviewed referred to the "Exxon" project. "In part, this is due to the fact that the partnership which formed CMC followed on an earlier all-Exxon venture. But we also sense a deliberate effort by opponents to invoke the Exxon name, with all of the negative associations related to it from the Valdez experience."

Where were the academics whose research and views could legitimately "counter the several 'professors' whose anti-mine comments we read?" Whether those critical views were factual or not, CMC must begin to balance them with varied proponents.[10]

Mike Monte of the Pioneer Express was unimpressed with CMC's public relations. "They were so transparent. They stretched the truth too much....We used to laugh at them." But the company had a tough sell, said Monte. "They were selling a hole in the ground with a good chance of wrecking a premier river in the state. Tough selling job."[11]

In the 1980s, federal involvement with the Crandon project was minimal, but that changed after 1993. Government's new role frustrated the mining company. Federal cautionary remarks came from the Green Bay office of the United States Fish and Wildlife Service in mid-November 1994 when Janet Smith shocked CMC with her critique of the company's plans. Her department believed the mine would diminish Indian interests in exchange for benefits for the general public, and federal agencies could not subordinate Indian interests to other public purposes. "The proposed project is highly complex and potentially could have significant adverse impacts on fish and wildlife habitat, water quality, Indian trust resources and other Indian interests," wrote Smith. A company consultant thought Smith's letter was the "first sign" that the federal government was going to be far more involved and make the project "much more complicated."[12]

Up to November 1995, CMC understood that the U.S. Corps of Engineers (COE) out of St. Paul, Minnesota was going to draw up an Environmental Impact Statement and would follow the same schedule as the Wisconsin DNR. To avoid duplication, much of the material in COE's federal EIS would be lifted from the state EIS, or so the company thought. COE's responsibility at Crandon was to preserve wetlands, supervise trust resources of the tribes (including hunting and fishing rights on ceded territories), and protect the tribes cultural properties.

After a meeting with COE in February 1996, Jerry Goodrich wrote Don Cumming, "It now appears that the federal EIS will be delayed at least a year after the state EIS and that little or no material from the state EIS will be used in the federal EIS." "Partners cannot accept a one year delay," Cumming wrote in a memo to his file.[13]

The Corps' St. Paul District Cultural Resources Management team was comprised of archeologists, historians, and geographic information specialists. In its rather vague mandate, the Corps of Engineers was to maintain the tribes cultural resources, meaning "the sum of the human experience which has significance to us either through the physical presence of a place or thing, or through an association of ideas or practices." It included prehistoric and historic period archeological sites, buildings and structures, and written records. Ceremonies, hunting practices, and social activities performed in order to maintain social or ethnic boundaries were also cultural resources.

The COE's separate study was done at the request of the Wisconsin tribes who didn't trust the Wisconsin DNR to do a fair one. "The tribes were concerned that the state would not adequately address impacts to trust and cultural resources," recalled Jon Ahlness, manager of the Crandon project for COE.

COE's decision, wrote Goodrich, would cause duplication in the environmental review process, delay the process, and ignore results based on generally accepted scientific methods, and force the company to redo rather than verify "acceptable work."

Over a nine year period, COE never completed its Crandon studies. Sissel Johannessen, who studied the cultural properties of Mole Lake and the Menominee for COE, thought COE had "a tremendous [bureaucratic problem]. It took ages. It was terrible. There was a blizzard of paper."[14]

In early August 1996, CMC received a copy of a twenty-page letter and sixty pages of accompanying documents sent by the EPA's Region 5 office in Chicago. The EPA letter criticized the environmental impact statement filed by the company in 1995. It attacked the reliability of much of the data in the company's EIS and insisted that many studies be redone and demanded a huge amount of new data and studies. "The August 1996 letter went off like a bombshell," said an Exxon consultant. Emergency meetings followed. At one meeting an EPA official alarmed the company's delegation saying, "Your client is still going to be trying to get permits even by the turn of the century."

"Yielding to EPA's demands would extend the permitting schedule by years and add untold millions of dollars to the permitting budget," CMC's lawyer, Chuck Curtis, advised. "EPA has become, for all intents and purposes, the federal advocate for the tribes. It has adopted the tribes' arguments hook, line, and sinker. It has abandoned all pretense of neutral decision-making. A hard-line, radical 'eco-green' bias permeates this entire document."[15]

The EPA posed several questions about the company's long-term financial responsibility. "It suggests, contrary to the DNR's repeated statements, that Wisconsin's perpetual liability law 'may have loopholes,'" noted Curtis.

"At this rate, we are not going to have a draft federal EIS next year or a final federal approval in 1998," wrote Curtis. The EPA letter could only be characterized as a "declaration of war" against CMC and could "effectively kill this project."

Curtis described another demand:

EPA insists that CMC examine compliance with the Sokaogon's new tribal water quality standards - the standards that prohibit anyone from doing anything anywhere that might affect a single molecule of reservation water or offend the Sokaogons' beliefs concerning the "sacred purity" of that water....Our position is that the Sokaogon have no sovereignty, ownership, or regulatory jurisdiction over the waters in issue and that, even if they did, the standards are blatantly illegal and unconstitutional.

Eventually the courts rejected Curtis' legal argument, and the EPA also insisted that CMC hire a cultural anthropologist and a social psychologist to study the "social-psychological health, belief systems, and sense of identity" of the tribes.

Finally the EPA demanded that the company redraft its EIS to include a sweeping analysis of the potential impact of the mine on other mining projects, for example, in northern Wisconsin and northern Michigan.[16]

"Neither state nor federal law requires such global examinations of the mining industry, or any other industry," Peter Theo, government affairs director for CMC, said in the company's public response. The EPA letter was based on "inaccurate assumptions, errors in fact and serious misinterpretations of, or confusion about state and federal law, which leads me to conclude that they could not have read all of the documents CMC submitted."

Daniel Cozza, the EPA's project manager for the proposed mine, replied that Crandon Mining had a duty to look at the cumulative economic and environmental impacts of other mining projects. If ore from the Crandon mine were smelted at the Copper Range Company near White Pine, Michigan, for example, that would affect the region's economy.

"Let's look at northern Wisconsin. What else might happen that would affect the economies of communities there, and the environment?" Cozza asked. "We didn't ask them to look into stuff that's happening in China or Australia."[17]

EPA's letter caused consternation within CMC. "The current strategy is to meet with the Governor, key State Legislators and select Wisconsin Congressional Delegates to discuss our options," said an internal company memo. "Chuck Curtis has informed CMC that this decision by the EPA has the potential to stop the CMC project."[18]

CMC also met additional frustration when attempting to establish a dialogue with the Potawatomi and Menominee tribes and particularly with the Sokaogon tribe at Mole Lake. "We had experts from all across the state come and tell us exactly how to deal with the native people and none of it seemed to work," Dick Diotte recalled. "We made very little progress with the natives." Had the Indians supported the mine, "I think [the permitting] would have gone through." CMC needed to study archaeological and socioeconomic problems and environmental issues, but the Mole Lake tribe wouldn't cooperate. The company did have one face-to-face meeting with Chairman Arlyn Ackley and the Sokaogon Tribal Council on May 31, 1994, but it was the only meeting Mole Lake representatives would attend.

Unlike his informal personal approach with Herb Buettner, Goodrich's communications with the tribes were formal and punctilious. "The tendency for Exxon was to send certified letters to the chair of tribes," observed an Exxon consultant. In a four-page letter, Goodrich informed the DNR's Bill Tans of the numerous failed attempts to communicate with the Sokaogon:

11/4/93   Letter to Chairman Ackley requesting assistance in data collection on threatened and endangered species. Also offered to discuss other topics that he wished to cover.   No response....

1/28/94   Letter to Chairman Ackley asking for meeting. Assistance in data collection: surface water sampling and survey of existing threatened and endangered species.
Sent certified mail - no response

2/17/94   Two CMC staff went to Mole Lake tribal office to seek permission to allow representatives of the environmental contractor to do aviary, mammal, soil and water quality testing on the reservation. Permission was denied on the spot....

8/12/94   Letter to Mole Lake Chairman Ackley reiterating requests lodged in letters of 6/2 and 7/15 and asked for a reply. Sent certified mail - no response.[19]

One Mole Lake tribal official, J. C. Landru, talked privately with the company, and some tribal members subsequently viewed him as a spy for the company.

J. Wiley Bragg was the company's contact with Landru who at one time had been Mole Lake's treasurer. "From several sources, we are told we are approaching Native American tribes inappropriately," Bragg wrote Landru on February 3, 1994. "We do not comprehend the cultural differences. Your assignment is to share with us pertinent information about their historical and cultural protocols and customs so we may better understand Native Americans and approach them with the respect they wish." Bragg wanted information about the tribe's religion and spirituality. "I know Mother Earth is sacred, but are there ways to honor this and still mine?"

"Please share with us names and a brief profile of tribal members, leaders as well as people working for the tribe that you feel is important for us to know," Bragg requested. He needed to know why did the Indian leadership oppose the Crandon Mine? Who among the Native Americans or environmental groups were the key leaders? What was their motivation? "We want your suggestions on meeting and addressing these 'anti-efforts.' Are you aware of any planned events against the Crandon project?"[20]

When Jerry Goodrich decided to retire in 1996, Rodney Harrill replaced him as president of the Crandon Mining Company. Harrill held a bachelor's degree in civil engineering from the University of Maryland and two master's degrees - in business administration and in civil engineering - from the University of Houston. Prior to joining the Crandon Mining Company, he had been a marketing manager for Exxon Coal and Minerals Company.

Harrill had spent most of his career doing "tangible and scientific" work. "You look at the facts and the evidence," he said. But Crandon was different. While there was science, engineering, and the permitting process, there were also public affairs and heated political controversies that were "quite different for me personally." Politics and public affairs unexpectedly consumed an inordinate amount of his time.

"I'd never been involved in a debate where politics and pseudo-science had the same weight as facts and scientific evidence," Harrill recalled. Getting a permit for the mine was "like the concept of infinity," he discovered. "In the year-and-a-half I was there, in spite of very focused effort on our part, we were no closer to getting a permit than when I started. It seemed we were always two years away."[21]

Harrill thought the Mole Lake Sokaogon were exceptionally disorganized. "I didn't view them as sophisticated enough to have a meaningful dialogue with." Harrill was referring to the chaos caused by Mole Lake's chairman, Aryln Ackley. Ackley was a natural leader, a dynamic speaker, and conscientious about most tribal issues, but he suffered from severe personal problems. He had been convicted in 1989 on four counts of selling cocaine and served two years in prison. Subsequently he was twice charged with drunken driving and wife battering. Then in 1997, while Rodney Harrill was president of CMC, Ackley was accused of embezzling from the tribe's gambling business. Later convicted, he served more time in prison.[22]

Harrill at least had dialogue with the chairman of the Menominee tribe, the charismatic Apesanahkwat. He met with Harrill eight times, usually to discuss possible job opportunities for the Menominee if the mine was constructed. "I gained a high regard for [him]," said Harrill. But at one meeting Apesanahkwat made clear that there would be no concessions to the white man. He opposed the mine "on principle." "Rodney, you must understand," he said. "We are against this mine just like we were against the white man coming to this country and taking over our land; just like we were against the white man putting us on a reservation; just like we were against the white man buying up all the land along the Wolf River and dominating the culture here."[23]

*   *   *

In 1995 Wisconsin's legislature eliminated the office of the public intervener. Governor Thompson argued that eliminating the office would save the state about $350,000 over two years. Environmental groups were powerful enough to hire their own attorneys to act as watchdogs. There was no need "for the state to pay attorneys to file lawsuits against itself."

The intervener's office had played a key role in fighting for environmental causes like the Crandon mine, a toxic landfill on the Fox River, and a shopping mall in Appleton. Local governments, frustrated by legal red tape, could turn to the public intervener to guide them through the bureaucracy.[24]

"No one claims the public intervener sues too often - in only 15 matters in the past 17 years!" a supporter commented. "The public intervener instead spends most of its time negotiating to find 'win-win' solutions to problems. And, no one claims that the public intervener sues frivolously - in spite of the fact that it takes on powerful agencies, the office usually wins." (It won a case in the U.S. Supreme Court.)

The intervener had been playing an important role in the Crandon mine controversy. The intervener had hired scientists and engineers to independently study the mining company's complex scientific models, and the consultants provided assurances that "what the company said would happen would indeed happen." The Crandon mining project was so huge that the intervener's office hired a attorney to concentrate on the mine's permitting process.

Critics questioned Governor Thompson's timing regarding elimination of the office of the public intervener. The intervener's office was focusing on researching and lining up expert testimony on the Crandon mine. Opponents were outraged about losing an agency that had assisted them so effectively. CMC also appreciated the public intervener because it served as a "buffer" between the company and hostile opponents. "I miss the intervener too," DNR's Bill Tans commented when asked about the grumbling by environmentalists. "I wish they had not been eliminated. They played a useful role for this agency."[25]

Public Intervener Kathleen Falk didn't know the exact reason Thompson eliminated the office, but speculated "It could have been my highway litigation. I beat the [Wisconsin] Dept. of Transportation on how it spent two billion dollars." She won in court. "I suspect [Thompson] wasn't pleased."

When the intervener's office closed, it ended a 30-year history as the public's environmental watchdog. "It was such a loss for citizens of Wisconsin," Falk lamented. "It had served citizens so well for thirty years. It was so sad. Citizens' groups don't have the resources, and individuals frequently don't have the resources."[26]

After the Crandon mine proposal's rebirth in 1993, George Meyer, the new Secretary of the Wisconsin Department of Natural Resources, gathered his top staff and told them to make decisions based on "your professional judgment, and let the chips fall where they may." In 1993 the DNR was far more experienced in mining regulation than fifteen years earlier. "We got geared up quickly," said the department's Larry Lynch, a hydro-geologist in the mine reclamation unit. Despite gearing up quickly, the DNR's permitting process dragged on for years.

Until 1995 the governor appointed the DNR board members, and they appointed the DNR secretary. Public faith in the DNR depended on the department's reputation for independence and honesty. Governor Thompson succeeded in elevating the DNR secretary to a position in the governor's cabinet. This action seemed to seriously compromise the agency's autonomy and reputation.[27]

Opponents of the Crandon project assumed that the pro-mining governor would order the DNR secretary to favor the mining company. Democratic State Representative Spencer Black of Madison claimed that the DNR had been subject to extreme political pressure since Governor Thompson assumed the authority to name the agency's chief. "The politically controlled DNR seems to be going out of its way in a manner that benefits the mining company." Even Jerry Goodrich privately referred to the DNR as an "ally."

Bill Tans thought the truth would come out in the future. "You don't have to trust us, or trust the company. You have to trust the system." But opponents of the mine didn't trust the DNR, the company or the system. "My confidence in the DNR is zero," said George Rock.

"There was a problem [with trust of the DNR]," Larry Lynch later mused. He thought the main reason was that the DNR never said the mine was a poor project. Instead the department tried to do an unbiased, scientific study based on rules and statutes. "We have to operate within the laws and rules."

"People had a perception that we were working with the company to make the project approvable and to some extent we [were]," said Lynch. "That put us at odds with people who were dead set against it."[28]

Meyer, the DNR secretary, claimed he had "good" relations with Governor Thompson. He told the governor that "When it came to regulatory decisions or enforcement decisions, those were [DNR] agency decisions and not the governor's decisions." Thompson agreed. "He never interfered with any regulatory decision or enforcement decisions during the time I served as Secretary," said Meyer who served from 1993-2001. "I don't think [Meyer] had a great relationship with the governor, but it was good enough to keep him in his job," commented the DNR's Stan Druckenmiller.

The project involved exceptional technical problems and challenges that were difficult for the public to fully understand. In 1995, Don Moe of the company wrote to the DNR's Chris Carlson about "Bedrock Layering:"

The regional numerical groundwater model currently consists of four layers that generally correspond to 1) the Late Wisconsinan Till and lacustrine deposits underlying Little Sand, Deep Hole, Duck and Skunk Lakes, 2) the coarse outwash, 3) the fine outwash, and 4) the Pre- to Early Wisconsinan Till. Due to interfingering of hydrostratigraphic layers, each model layer may contain portions of two or more hydrostratigraphic layers. A weighting scheme, arithmetic averaging for horizontal hydraulic conductivities and harmonic averaging for vertical conductivities, has been used to account for changing hydrostraigraphic properties within model layers.[29]

Initially, CMC had assumed that they could easily get a permit because in the 1980s they had almost completed the process. Now they expected to obtain a permit decision in three years. "That affected the dynamics of the whole project at the outset," said DNR's Carlson. It created "antagonism." In 1994 on Carlson's first trip to the Crandon site, he met Don Moe. While talking about the permitting process, Moe became enraged. "We were standing by Swamp Creek, and he just turned on me and just lit into me, and said, 'You are out of your mind! You are going way beyond what is reasonable!'...His face was just bright red. He was just ripping into me," Carlson reported.

Carlson didn't think that CMC effectively explained the project, that it couldn't deliver a coherent believable message. "They [underestimated] the nature of the problem." A CMC consultant in Vancouver, for example, performed a lab study that could not be replicated by the DNR, according to Carlson. "It was not validated or verified anywhere." Foth and Van Dyke seemed unable to handle a project as complex as Crandon, in Carlson's judgment. The company was "ticked at us, but we were at least as ticked at them for giving us crappy work...that was not as well thought out as it should have been." The DNR's mistrust of CMC's submissions delayed the permitting process.[30]

In 1995 Larry Lynch investigated whether a metallic mine had ever been built, operated, and closed with no problems. He couldn't find such an example, a position that emboldened mining opponents who wanted a moratorium on mining. Despite not having any history of proven technology in the mining industry, Lynch argued in the same report that there was "a great deal of experience with such measures in other waste management fields," which eventually could be applied to mining, a position that angered opponents and led them to question the objectivity of the DNR.[31]

When the company completed development of its groundwater model, the DNR obtained the model software and performed checks and analyses of the modeling effort. The goal was to "verify inputs, calibration, sensitivity analyses, boundary conditions and other characteristics that would affect the usefulness of the model's predictions."

In July 1996, several legislators petitioned the Natural Resources Board, requesting that the mining rules be changed to require a mining company to obtain insurance to cover costs of remedial actions in the case of serious accidents or spills. The petition also asked for a review of the highly controversial groundwater protection rule granting the mine a compliance boundary distance of 1,200 feet.

The DNR endorsed the revisions and suggested an irrevocable trust fund that the applicant would have to fund before mining could begin. Although a mining company was already liable forever for environmental problems, the DNR said, "the fund would provide a guaranteed backup source of funding." Both revisions were implemented.[32]

In early 1997, because of the growing avalanche of criticism against the DNR, the department scheduled a series of public information meetings around the state. "Normally our EIS process provides for adequate public input to the process," Bill Tans explained to the DNR's staff. "This is not an ordinary project, and the public needs to hear from us before the public meeting on the draft EIS....Because of the significance of the public debate on Crandon project issues, I see no substitute course of action."

"I have thick skin," Meyer said of the criticism he received from company officials and from zealous opponents of the mine. He publicly replied to noteworthy criticism, he said, to defend DNR staff, particularly from the charge of political influence. In a standard reply delivered hundreds of times, Meyer assured a critic that the state's mining laws were strict and comprehensive. "We could not permit a project if it would make groundwater undrinkable, for example, or pollute surface waters and kill even the most sensitive aquatic organisms."[33]

Stan Druckenmiller admired Meyer's integrity, but working as his executive assistant was not always pleasant. "He was a moody person," Druckenmiller said. "At times he did not treat staff as he should have. George wanted us to respond in kind to every stupid news release by the opposition," said Druckenmiller, who judged the policy too time consuming.[34]

The DNR also tried to reach out to the tribes but wasn't always successful. "We were left with the specter of ultimately going into the [master] hearing and getting to a permit decision without substantial tribal input," said Druckenmiller. "They just didn't want the mine." However the Menominee tribe respected George Meyer's integrity and good intentions. "We knew that he had a strength about him," said Ken Fish. On the science of the mine when there were challenges, the DNR "went deep into verifying those mining technologies," and took the Menominee's comments "into consideration." "Even though it appeared on the surface that he was pro-mining, there was an underlying commitment that the science be looked at in detail," said Fish.[35]

The DNR sometimes deserved censure because of poor public relations, and many felt it paid scant attention to past performances of mining companies. Wisconsin law required the department to consider a mining company's track record, said the DNR, but only within the United States within the previous ten years and only for mining-related projects. Other environmental incidents, such as the Exxon Valdez, "would not have any relevance in the decision process."

Because the DNR never clarified why Crandon Mining Company had been given a 1,200 foot compliance boundary, whereas all other waste dumps in the state had a 150 foot boundary, opponents hollered "favoritism." "Our concern is that today's mining laws allow a 1,200 foot compliance boundary which the mining company can legally contaminate with their toxins and contaminants," Sonny Wreczycki told the Natural Resources Board in 1996. "This allows over two square miles of contamination to our drinking water. There is no protection for the people's drinking water to remain as it is today." The boundary disparity was eventually changed, but only after fifteen years of complaints.[36]

Opponents worried that the DNR had the discretion to waive laws and grant variances for mining when the rules were too strict. For example, the Kennecott copper mine project next to the Flambeau River was permitted within 140 feet of the river, instead of the standard 300 feet. Dave Blouin told the DNR's Larry Lynch that the perception among the department's critics was that the DNR "never openly criticizes anything Exxon does."

"If the DNR is so impotent it can only make rules to satisfy multinational corporations, it is indeed a neutered cat and of no use to the citizens of Wisconsin, and actually, is a detriment to them," Roscoe Churchill wrote Meyer.[37]

One critic deconstructed a DNR brochure that focused on twenty-five public "misconceptions" about mining in Wisconsin. It inferred only opponents of the mine were guilty of misconceptions. "Basically, you have your misconceptions confused," said the critic:

To believe that the world is flat is a misconception. To believe that mines can pollute or that there could be a draw-down of water in the Wolf River - those are not misconceptions. Those are real possibilities...You've got a real public relations problem and you're certainly not going to address it by putting out a piece as condescending as this, which is predicated on the assumption that the public is disillusioned or misguided (translated "dumb"). You end practically every argument by saying in essence that the DNR will not allow whatever calamity is being addressed, to actually happen. That's the crux of the issue. But don't be patronizing and tell us all of our fears are predicated upon false conceptions.[38]

When challenged on this imbalance, Meyer replied in a radio interview that he didn't know of any misconceptions promoted by the mining company. He agreed that the brochure might have ignored company misconceptions. "[I] agree with your comment that our recent publication, "Misconceptions About Mining In Wisconsin, does not cover many of the issues you included...I will ask staff to consider some of the specific concerns you listed." But the DNR never published a more balanced version.

Meyer claimed he never saw evidence that the DNR "dragged [the permitting process] out because we didn't want to make a decision." Others thought differently. The DNR "was in a lose-lose situation," commented company engineer Gordon Reid. "The only way they could win was to not make a decision. Any decision the DNR made would be criticized. If they said the mine would be fine, they would have been criticized as corrupt. If they said, no, there was a fatal flaw, we would have sued them."

"That (timing) would probably be my biggest criticism of our process," said Larry Lynch. "It does go on so long without specific milestones or specific timeframes in which we have to operate. We allow things to drag on."[39]

An unsettling moment in George Meyer's career came in the winter of 1999 over lunch in a Madison restaurant. There Chuck Chvala, Democratic Senate Majority Leader, had a proposition for him. Earlier when Governor Thompson had reappointed Meyer as secretary of the DNR, the Senate's Natural Resources Committee voted 5-0 for his confirmation. While the full Senate confirmed other cabinet members, Chvala blocked Meyer's confirmation. Over lunch Chvala told Meyer, "You will not be confirmed. I will not bring you up for confirmation, unless you guarantee to me that the permit for the Crandon Mine will be denied." "Chuck, I can't do that," Meyer responded. "There is no way legally that I can do that...It is not proper." Meyer was never confirmed, but he served anyway because the Governor could fill vacancies.[40]

*　　*　　*

In March 1995 Crandon Mining announced a change in plans that would radically alter the political prospects of the mine. Since the company would not be allowed to discharge wastewater into the protected Wolf River, it had to come up with an alternative. It decided to discharge its wastewater through a 38-mile pipeline to the Wisconsin River.

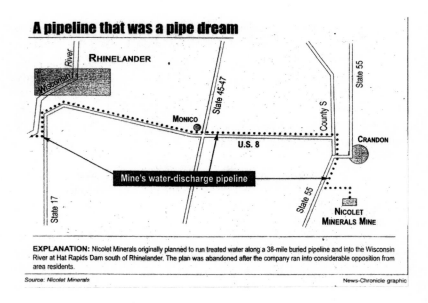

A pipeline that was a pipe dream

EXPLANATION: Nicolet Minerals originally planned to run treated water along a 38-mile buried pipeline and into the Wisconsin River at Hat Rapids Dam south of Rhinelander. The plan was abandoned after the company ran into considerable opposition from area residents.

Source: Nicolet Minerals

News-Chronicle graphic

Goodrich had accepted the advice of his engineers and his management committee, and two weeks before he publicly announced the pipeline decision, he met with advisors and the company lobbyist, Jim Wimmer, and a public relations consultant, Jim Wood. At the meeting Goodrich outlined three options for controlling the mine's wastewater: Send it to the Wolf River, but the Wolf's designation as an Outstanding Resource Water made that impossible; manage it on site, but there was too much water; or dump it into the Wisconsin River. Goodrich saw no other choice than the Wisconsin River. The 38 mile pipeline would travel west along Highway 8 to Rhinelander, then south along Highway 17 before dumping the water into the turbine at Hat Rapids, about one mile north of the Lincoln County line.

Wimmer spoke eloquently against the move: "You begin to discharge water into the Wisconsin River - and I don't care how clean it is - you're going to create a political issue all up and down the Wisconsin River!....It's going to explode into a statewide environmental issue." Wood agreed. "I wasn't happy with the decision," he recalled. But Goodrich saw no alternative.[41]

Years earlier the water quality of the Wisconsin River was much worse than in 1995. Back then the wood pulp and paper industries and municipalities had provided only rudimentary treatment of wastewater effluents. The main problem in the river was the depletion of oxygen. Without oxygen, fish could not survive. The effort to improve the river focused on controlling the oxygen demanding substances (biochemical oxygen demand - BOD) in the effluents. In the 1970s and 1980s, with the passage of federal and state environmental legislation, the industries and municipalities were forced to install treatment systems for their wastewater - primarily to reduce the BOD. The result was a significant improvement in the quality of the water, but the waterway still had dozens of industries and municipalities dumping wastewater into it since the DNR had established more relaxed limits on discharging than it allowed for the Wolf River.

Several municipalities and industries were already discharging into the site preferred by the Crandon Mining Company, and the DNR argued that any new or increased discharge into the waters of Wisconsin could not be prohibited, provided that the discharge met the state's water quality standards. Nor would the DNR dictate where a company should discharge its wastewater. "If a proposed discharge can meet the water quality limitations, it must be permitted. And if it doesn't, the discharge would be prohibited."[42]

Initially the DNR's position on the company's plan was that the river could tolerate the "modest" increase of BOD levels caused by the company's pipeline discharge. In November 1996, Bill Tans claimed that the pollutants discharged into the river would be sulfates, dissolved solids, and several metals "in very, very small quantities....There will be very little effect on the river water."

The company and the DNR agreed that the discharge would be water that had flowed into the mine, been treated in a water treatment facility, tested for compliance with the effluent limits, and then sent through the pipeline at the rate of a million gallons a day. "While the figure of one million [gallons] per day in isolation may seem like a large volume, in terms of the flow of the Wisconsin River it is relatively small," Larry Lynch contended. "The average annual flow of the river at the proposed discharge site is about 486 million gallons per day."[43]

Most people in western Wisconsin viewed the situation differently than did the DNR and the company. Until then the

Crandon mining controversy had seemed remote. Crandon was just a "little town way up north." But the pipeline plan broadened the Crandon Mine controversy into more neighborhoods, and residents resented the company's audacity. A multinational mining company wanted to build a mine in one county (Forest), ship their waste across another county (Oneida), and dump it into a third county (Lincoln). The DNR lost credibility by neglecting to notify Lincoln County of the pipeline plan, apparently naively believing the wastewater discharge would not offend the people who used the river every day.[44]

The first place pollutants would settle was in a heavily populated 1,400 acre reservoir, about nine miles below Hat Rapids, called Lake Alice. After learning about the pipeline, Jim Wise, an environmentalist and businessman in Tomahawk, helped form POWR, Protect Our Wisconsin River (not to be confused with POW'R in Shawano). On February 12, 1996, Wise spoke at a Town of King meeting, a short distance from Lake Alice. As he summed up the pipeline plan, he recalled, "There were a lot of mouths hanging open." None of the town board members and no one in the audience knew about it.

Later, when the mine company's vice president, Ken Collison, spoke to eighty-five people at the Town of King, he made a major public relations blunder. A member of the audience asked, "What's going to happen to Lake Alice?" "Lake Alice doesn't mean anything to me," Collison responded. "Where's Lake Alice?" Subsequently critics distributed thousands of bumper stickers that read, "Where's Lake Alice?"[45]

"We were outraged," Wise recalled. "We thought it was very unfair that the system allowed for a company to dump their waste so far from the source of their economic benefits to a county that benefited not one bit from the waste. We would get nothing but the waste."

In addition, opponents of the pipeline didn't believe the contention that the pipeline's wastewater would be harmless. They feared it would further pollute the Wisconsin River and set a precedent for other industries to pump wastewater long distances to avoid strict pollution standards on other waterways. "We spent thirty years cleaning up the Wisconsin River," said Wise. "Why are we now looking the other way when another company starts to re-pollute the Wisconsin River?"

In January 1996, an environmental group asked the DNR's George Meyer to come to Lincoln County to explain the pipeline proposal, but he refused the invitation. Two months later, amid protests, he agreed to come. At a five-hour public hearing on the pipeline at Tomahawk High School, fifty pipeline opponents testified. No supporters rose to speak. "At the beginning of the meeting," reported the Wisconsin State Journal, "DNR officials conceded that public opposition had surprised them and that anti-mining activists had forced the public hearing." "Initially we decided not to hold a hearing," George Meyer explained. "But give credit to those who were vocal and persistent. More and more people told us that this was a major change, and that's why we're here tonight."[47]

One person who agreed it was a major change was Bob Kranda, who wore a "Where's Lake Alice?" button. Kranda and his wife were among thousands of ordinary citizens who became anti-pipeline activists. Large windows in the Krandas' home provided a view of deer and birds on their lawn and Lake Alice lapping its edges. They had moved to Tomahawk in search of clean water, clean air and wildlife. "Their children and grandchildren visit regularly for swims, boating and fishing," noted a reporter. "'If the Wisconsin became polluted again, what will they have to leave to their grandchildren?'" the Krandas asked.

Both Krandas had retired two years earlier, but the pipeline issue kept them busy. "I'm on the phone all day," Bob said. "I was never an environmentalist before, but the more a person finds out, the more a person thinks he'd better get involved. My son-in-law said 'You were once a nice guy 'till you got so radical.'" Kranda put his hand to his forehead and laughed. "We'd better get involved, because you can't trust the government to watch the public's best interests," he said, "especially since the DNR secretary is appointed now by the governor....They can't go against their boss, and Gov. Thompson said he wants mining in Wisconsin."[48]

POWR countered CMC's public relations campaign with press releases, yard signs, bumper stickers, letters to newspapers, forums, and lectures in schools. Bart Olson, publisher of the Shopper Stopper in Merrimac, joined the anti-mining movement because of the pipeline. "I was upset. I live on the [Wisconsin] River. It must be horrible stuff if they couldn't dispose of the [waste] near the mine. If it is treated water [as they said], let it seep back into the groundwater over in Crandon." The Shopper Stopper ran front page editorials protesting the pipeline.

A Sauk County board member, Bart Olson convinced his board to pass a resolution against the pipeline and asked nearby municipalities to do the same. "By the time I was done I had thirty," Olson said.

Due to the concern of his constituents along the Wisconsin River, Representative Eugene Hahn of Cambria became the first Republican to break party ranks by introducing legislation to ban the company from dumping wastewater into the Wisconsin River.[49]

While the pipeline proposal roiled western Wisconsin citizens, it was also disputed by some of the leading political leaders in the Midwest. United States Senator John Glenn of Ohio and Michigan's Governor John Engler led a group of a dozen political figures to protest the pipeline proposal since the federal Water Resources Development Act of 1986 set up a protocol of state approval for all diversions of water outside the Great Lakes Basin. The CMC pipeline would draw groundwater from the Great Lakes Basin for use in its mining processes and discharge the treated wastewater into the Wisconsin River, which resided in the Mississippi River Basin.

In his protest to Governor Thompson, Governor Engler stated that the proposal set a dangerous precedent and "presents important legal and public policy issues, not only in Michigan, but to all the Great Lakes States." The Water Resources Development Act prohibited a diversion of any portion of the Great Lakes unless such diversion had the "unanimous gubernatorial consent by all Great Lakes State Governors."

Thompson and the state DNR argued that the law said diversions between two and five million gallons required a permit from the DNR. Only those diversions in excess of five million gallons per day required the agreement of the other Great Lakes Governors. "Because the proposed amount is less than two million gallons per day, under our law the mining company's withdrawal does not require a permit from the State of Wisconsin nor do we believe the concurrence of other Great Lakes States."

The Wisconsin DNR also noted that the federal laws referred to diversions of water from the Great Lakes or its tributaries, but did not specifically include groundwater diversions or withdrawal. For that reason, said the DNR, "our interpretation of the 1986 federal law is that it does not apply to the proposed mine [pipeline]."[50]

"Groundwater often interacts directly with surface water making a hydrological distinction between the two of little practical merit," John Glenn retorted. "Clearly, the diversion of water from a groundwater source that communicates with a surface stream or lake has the effect of diverting surface water. Such an action should be subject to the approval of the eight Great Lakes Governors."

In the end, the U.S. Army Corps of Engineers in St. Paul decided in favor of Wisconsin's position. The transfer of water from the Great Lakes Basin as proposed by Crandon Mining would not be illegal.[51]

By then the issue was moot because pressure by opponents in western Wisconsin caused the DNR to reassess its tolerance of the company's pipeline plans. "Due to the controversy over Crandon Mining Company's proposed discharge to the Wisconsin River, the perceived over-allocation of the river became a local issue," Bill Tans wrote George Meyer in May 1997. "The question being asked was 'How can you give to Crandon Mining Company an allocation for BOD when the river is already over-allocated?'" It was time for the DNR to restudy the problem and determine what the capacity of the river was to assimilate BOD.

The DNR's Stan Druckenmiller informed Rodney Harrill of the bad news. The department had decided that the summertime levels of BOD in the river were "fully allocated and over-allocated under some conditions." Consequently the department could "not approve any discharges of detectable amounts of BOD" until it completed the rule revision process. "This entire process may take two years or longer." Druckenmiller suggested that there were several other potential discharge options for the company if it did not want to wait two years. In effect, the DNR's reconsideration killed the pipeline. "The pipeline has backfired," Zoltan Grossman exulted.

"It's bringing in a whole new group of people who otherwise would not have been interested [in the anti-mining movement]."[52]

The company's tactics angered mine supporters. Now that he was Secretary of Administration under Governor Thompson, James Klauser, the former Exxon lobbyist, still sympathized with the project but questioned the company's judgment. "I think that had a devastating political effect on the project," he recalled. "Exxon and their partners never sold [the pipeline idea]. I talked to the Exxon folks. They came [to Madison] from Crandon. I told them 'I can't imagine why you did this with the ramifications without the Exxon process which had always been very deliberate.' Almost too deliberate in thinking things through. 'This is unlike Exxon to have made this major change without understanding the ramifications.'"

"I was direct and harsh," Klauser recalled.[53]

✱   ✱   ✱

In mid-1996 Mary Ann Pires suddenly noticed a downturn in the company's public support. "I spoke with four people in Crandon today," she wrote Goodrich in June. "My sense that CMC community support is being rapidly eroded was confirmed in EVERY CASE."

Dale Alberts, the company's chief lobbyist in Madison, noticed the decline in public support at about the same time. On July 26, 1996, he warned of the company's dire predicament:

The government and public affairs program associated with the Crandon Project is in disarray. It has been fragmented and largely ineffective. Political and public support for the project is eroding rapidly.

The Crandon Project is in serious and immediate jeopardy from the eroding political support and subsequent possibility that the legislature may pass a moratorium on sulfide mining during the 1997 session.

An aggressive and well focused program designed to win back public and political support must be implemented immediately.

Public relations consultants had arranged multiple programs designed to win public support and "yet the programs have not been fully or properly implemented," Alberts wrote.[54]

The negative reaction to the pipeline was one reason for the decline in public support for the company. More important were the efforts of the Wisconsin anti-mining movement.

CHAPTER SIX

"SAVE THE WOLF RIVER!"

Shortly after Jerry Goodrich's press conference in 1993 announcing the revitalized mining project, two events revealed increased opposition to the mine.

The <u>Rhinelander Daily News</u> ran a full-page advertisement, signed by local businesses, welcoming the return of the Crandon Mining Company. The Mole Lake Sokaogon, incapable of effective protest in the 1980s, quickly announced a boycott requesting readers to refrain from purchasing products of the businesses who paid for the ad. The local distributor for Coca-Cola and a major food retailer immediately felt negative effects when they lost Mole Lake's casino contracts as did a construction firm that had anticipated doing roadwork for the tribe.

"These companies have already demonstrated their arrogance and disregard for the rights and well-being of our people in making this announcement," said Arlyn Ackley, the newest chairman of the Mole Lake Sokaogon. "Our tribe was not even notified that an announcement was to be made. This is outrageous."

J. Wiley Bragg, the company's public affairs specialist, privately explained to Jerry Goodrich that Dick Timmons of the <u>Rhinelander Daily News</u> simply thought he was making a good civic gesture. "The 34 sponsors agreed, like they probably have followed Timmons lead on numerous such ads over the past years. Everyone, including CMC, was shocked at the Mole Lake reaction."

Bragg continued, "Arlyn Ackley was offended by such a strong pro-mine stance since they are convinced mining cannot be done without major adverse impacts to the environment and their way of life. Ackley is seeking apologies from Timmons and the 34 ad sponsors."[1]

The second event occurred on Saturday, April 23, 1994, at the Nashville Town Hall during the first public hearing on the new mine proposal. The DNR's staff "got an earful." "We learned how things had changed," said Larry Lynch. The crowd was much larger and far more contentious than in the 1980s. Cars and trucks parked for blocks up and down roads near the town hall. Wrote one reporter, "Saturday evening, about 300 people attended the second hearing session and more than 100 voiced their opinions while about 15 DNR employees served cookies and coffee, and listened." The DNR officials remained past midnight and still couldn't accommodate all those who wanted to testify. Only a handful of people favored the project.

Mayor Vernon Kincaid was dismayed with the hostile audience at the public hearing. "It seems as though there was quite a lot of personal animosity shown towards individual people of the Crandon Mining Co.," he wrote afterwards to the DNR's Lynch. "It seems that the worse the statement made against mining officials, the louder the applause and the louder the beating of the Indian drum."[2]

What had happened to dramatically increase the efforts of opponents? The three tribes closest to the mine and several allied tribes, their financial position greatly enhanced by casino profits, adamantly and aggressively opposed the mine. For the most part, they would not negotiate nor engage in dialogue with the Crandon Mining Company representatives. Moreover, Indians joined with non-Indian groups, including former adversaries, to protect northern Wisconsin's natural resources.

The economy in northern Wisconsin, bolstered by jobs in casinos, had improved dramatically and undermined the company's argument for providing new long-term jobs. Mining disasters everywhere energized the expanding anti-mining movement. The Exxon Valdez remained on everyone's mind. Opponents worried about a similar environmental disaster at Crandon.

Sonny Wreczycki, George Rock, Herb Buettner and Al Gedicks resumed their anti-mining efforts, and they were joined for the first time by a score of other leaders and powerful organizations. Unlike the 1980s, the Sierra Club added resources into the fray. The recently formed Midwest Treaty Network sponsored a highly effective project that educated local communities about metallic sulfide mining. Local anti-mining organizations, like the hardworking spirited group in Shawano County, distributed literature and presented resolutions before municipal governments. Fishing groups, normally conservative and non-partisan, actively opposed the mine, and their efforts influenced state legislators.

Opponents became media savvy, using the Internet and creative web sites. Events were changing political power and relationships related to the mine and damaging the company's credibility.

National and state media now viewed the conflict as newsworthy. In late 1994 the <u>New York Times</u> described the Crandon project as "one of the country's fiercest grass-roots environmental face-offs." A Wisconsin newspaper agreed, predicting that the project "will be a massive, emotional, no-holds-barred battle to the finish." It was an "emotionally charged subject," added Bill Tans, mine project coordinator for the DNR. "It's going to be a bear of a [master] hearing."[3]

*   *   *

The tribes near the mine continued to worry about poisonous leachates from mine wastes. They believed the mine would poison their water, kill their fish, destroy their forests and ruin their tourist industry. "Ten years ago, Wisconsin's Indians had to persuade people to their cause on principle," noted the <u>Appleton Post-Crescent</u>. "Today, with millions to spend from casino proceeds,...the alliance can afford to map out an extensive strategy to halt the Exxon mine."

Four Indian tribes - Sokaogon, Potawatomi, Menominee and Stockbridge-Munsee - forged an alliance, the Niiwin Intertribal Mining Council, that promised a formidable fight against the mine. At a private meeting Menominee leader, Louis Hawpetoss, implored fellow Indians in other tribes to stay united and firm. "If you open the door for a monster like Exxon, you're putting a split into the unified position....As a Menominee, I cannot afford to have anyone present a plan like this. It is in total conflict with what I believe."[4]

The Council worked closely with the Midwest Treaty Network, Wolf River Watershed Alliance, Trout Unlimited, Wisconsin Resources Protection Council, and the Environmentally Concerned Citizens of Lakeland Areas. "This alliance has the opposite effect of the spear fishing crisis a few years back," Al Gedicks accurately observed. "Instead of Indians and non-Indians being in conflict over Northern resources, they have come together to protect the same resources from an outside threat." The futuristic language used by white organizers amazed Hawpetoss. "'Seventh

Generation' just rolls off their tongues. I've never heard non-Native people talk like that."

In the past the Menominee tribe had endured exceptional trauma. In 1954, the United States Congress passed the Termination Act, an experiment to force tribes to join the mainstream of American society. It went into effect in 1961. The federal government judged the Menominee tribe self-sufficient, progressive, an ideal candidate for assimilation. The experiment was a disaster. Menominee County, formed as a result of the misguided termination effort, became Wisconsin's 72nd and poorest county. The tribe lost land and assets and was restrained in its self-determination.

Because of the failure of termination, Congress reversed its decision. The historic Menominee Restoration Act of 1973 granted the Menominee status as a sovereign Indian nation to which the federal government was obligated by treaties, agreements and statutes. Tribal government was reestablished in 1979.

Regarding the proposed mine, the Menominee Indians were firm in their conviction that it would be detrimental to the Wolf River and the economy of the area. "We ask that this permit process be halted immediately," they demanded of DNR's George Meyer in 1995.[5]

Ken Fish, the Director of the Menominee Treaty Rights and Mining Impact Office, became a spokesman for the three tribes nearest the mine site. Testifying before the state's Natural Resources Board in 1996, Fish said, "Would you today contaminate your Mother?....We have to protect [clean water] at whatever stakes. I'm not here to bring you all kinds of technical data. I'm here to appeal to your conscience, your conscience for looking towards the future. When we as Native Americans look to the future we look at seven generations....Since the first treaty was signed in 1817, we still maintain a clean healthy Indian reservation free of pollution....We pride ourselves on the clean water that flows through our reservation and that comes up through our natural springs on the reservation....You cannot even put a price tag on what the Menominee Indian Reservation is worth today."[6]

At a hearing with the Corps of Engineers in 1995, several Menominee Indians wept openly as they told COE officials of their

112

reverence for the Wolf River. "Water flows," said Frances 64, of Neopit. "It flows into the earth. It flows underground. . cannot allow our water to become polluted. We are responsible for our children for ages to come. We cannot let them die from weird diseases." Others also spoke grimly. "They must have no spirit or soul to do this to Mother Earth," said Virgil Barnes. "This is the final insult. To me this is a declaration of war. It is not about to stop. Not when there is so much money involved."[7] — *Sparkey + dynamit for real*

When individual Menominees talked with company officials, company president, Don Cumming, found discussions cordial but anchored in firm beliefs. "They would sit there and listen to [me], and then look [me] in the eye and say, 'That's fine, Don, but we don't want the mine'." Apesanahkwat, chairman of the Menominee Nation, would not negotiate with the mining company, but he did discuss the project with company officials, thereby alienating a portion of his tribe. "He did that on his own," said Ken Fish. "It wasn't the tribal government's wishes. It wasn't good."[8]

Gus Frank, chairman of the Potawatomi, said, "Everybody had a fear of what the mine was going to do to us. It would have left a real threat to our environment in the North woods, to our water table, our drinking water, our recreational use, our rivers and our lakes." The company made promises, but Frank didn't believe them. "Show me. You can tell me something. But show me. Show me where you acted responsibly." The Potawatomi's communications with the company broke down in about 1994. "It got to be very adversarial," said Frank.[9]

In June 1994, several hundred Indians, some from as far away as Alaska and Colombia, convened for the "Protect Mother Earth" gathering at Mole Lake to oppose the Crandon mine. The conference was co-sponsored by the Indigenous Environmental Network and the Midwest Treaty Network. "This is to put Exxon and [Governor] Tommy Thompson on notice that we can bring people up here to stop the mine," said Bill Koenen of Mole Lake.

"It makes me happy to see a place like this. It's really pretty," said Don Many Bad Horses, chairman of the northern Cheyenne Nation. As a consequence of mining near his reservation, he commented, "Our water is ruined. We have to buy water now."

Round Table Meeting Mole Lake

Several reporters covered the event. "Sweat lodges were held on the Sokaogon grounds as Indians...gathered to develop strategy to stop the Exxon mine. The Potawatomi guarded the sacred fire, tobacco ceremonies were held and three ceremonial pipes were passed to be puffed by young and old. The smoke from smoldering sage was caressed as one leader told of the Indians' 'special connection with the Creator.'" Tribal elders and grandmothers enjoyed special places of honor. Teenagers drummed and chanted prayers to the Creator and Mother Earth.

Unlike other environmental rallies, Indian leaders opened theirs with spiritual messages, honoring their ancestors, their warriors, and their pledge to future generations. "We want to fight this mining in a spiritual way," Louis Hawpetoss said, "so we're ably guided by the spirits." The tribes always spoke from the heart, said Linda Sturnot. "They prayed at meetings. It was just a very powerful, powerful movement."

CMC publicly claimed it was committed to building a productive dialogue with the three tribes on matters of environmental, economic and cultural concerns, and the company pledged to respect the separate "sovereignty, culture, traditions, heritage and diversity" of each of the tribes. But the tribes didn't believe the company.[10]

Stress in the Mole Lake tribal office was palpable. "Stress from wondering if you could do enough," recalled Tina Van Zile who worked on environmental issues. She feared the company's lawyers and worried if she had sufficient expertise. She had to be alert and responsible every day. "What you do every single day could have affected the whole tribe."[11]

While Mole Lake's efforts aggravated many company officials, others appreciated the tribe's plight. "I understood the Sokaogon because the mine site was about two miles upstream from them," said Gordon Reid, an engineer. "If anyone would be impacted, they may have been impacted. I really respect and understand their having a concern. In their minds the only risk was no risk....The only way to ensure nothing would go wrong was not to have the mine."

The Mole Lake tribal chairman, Arlyn Ackley, told Jerry Goodrich that he wanted to avoid negotiating with the company. "I didn't want to give them any hope of meeting with my tribal council." The tribal council agreed. However at times Ackley was downright abrasive. He had told Wisconsin's Governor Earl, "We'll dig up your parents' graves, and see how you like it."[12]

"The [Mole Lake] Tribal leadership doesn't like the Exxon project, doesn't trust the state...agencies involved, and feels that the project is 'life threatening' to their traditional way of life on the Reservation," a DNR official wrote George Meyer. "I am not real optimistic about maintaining a viable and productive future relationship with this Tribe." Indeed, Mole Lake's leaders didn't trust the DNR. "I thought they were working together - Exxon and the DNR," said Arlyn Ackley.

The DNR's Chris Carlson knew the Potawatomi and the Menominee tribes were minimally engaged in the regulatory process and had technical staff and consultants who provided worthwhile feedback. But the Mole Lake tribe was Carlson's biggest headache. They refused to allow the DNR to have access to the reservation. "So we could not do any testing. They also refused to provide us [with the results] of their testing." Mole Lake threw up "road blocks with no engagement or alternatives suggested."[13]

Meanwhile, the DNR resented the involvement of the EPA and COE in the controversy, particularly their sympathy towards Mole Lake. Bill Tans said he had "all kinds of problems" with the federal intervention. The state already had standards to protect the environment for human health, welfare and safety. Did the federal government have standards that were higher? Would they protect the groundwater any differently than the DNR? "I can't see where they would go beyond where we would go," said Tans.

When the Clean Water Act of 1972 was amended by Congress in 1987, it granted tribes the right to enforce their own water standards. The EPA contended that Congress had authorized the EPA to treat Indian tribes the same as states. On September 29, 1995, the EPA granted the Mole Lake Sokaogon's application for Treatment as State (TAS). This gave the tribe the right to establish water quality standards for their portion of Swamp Creek and other tribal waters. The State of Wisconsin immediately sued the EPA in federal court, arguing that the agency had exceeded its authority since Mole Lake had lost control over reservation waters when Wisconsin was granted statehood in 1848. The state also questioned Mole Lake's ability to manage a water quality program. In 1999 the U.S. District Court in Milwaukee dismissed the suit, and the state immediately appealed. On September 21, 2001, the Seventh Circuit Court of Appeals in Chicago sided with the EPA and the Sokaogon. When the U.S. Supreme Court let the lower court's decision stand, finally the EPA and the Sokaogon had won. Mole Lake's high water standards meant that the mining company would have to return water from its mine to the same pristine quality as when it came into contact with the mine.[14]

Many citizens deeply respected the Mole Lake tribe for their tough principled stance. "A lot of people are coming to see the [Sokaogon Chippewa] as the real heroes here, holding their ground," said Mike Monte editor and publisher of the Pioneer Express, the Crandon weekly.

*   *   *

Past mining disasters throughout the country energized opposition to the Crandon mine. Devastation caused by old copper and iron mines littered Michigan's Upper Peninsula. Colorado, California, and Montana featured ghost towns, poisoned watersheds, and burial mounds of waste rock caused by mining long ago. The U.S. Bureau of Mines estimated mine wastes had contaminated 10,000 miles of rivers. Mine clean-up projects made up forty-eight of the United States Environmental Protection Agency's Superfund sites with the clean-up costs in the billions.

Some of the catastrophes were recent. In 1985 a small mining company, Galactic Resources, announced that it had discovered a huge gold deposit in Summitville, Colorado. Nine years later the mining company was bankrupt, leaving behind $20 million in unpaid bills and millions of gallons of contaminated water, some of which had spilled into the nearby Alamosa River. The EPA estimated the cleanup cost at $100 million. The Department of Interior Secretary Bruce Babbitt called it an "extraordinary disaster."[15]

Critics worried about an environmental catastrophe at Crandon. If there was an "overflowing toxic waste dump, a leaky pipeline, a cloud of poisonous tailings dust," the Crandon Mining Company could simply declare bankruptcy, forfeit minimal bonds and leave a multi-million-dollar Superfund clean-up to future Wisconsin taxpayers.

Wisconsinites should not be as gullible as the residents of Summitville. "We can learn to be persistent and question every facet of the Crandon mining operation - and not be so eager to swallow all those promises of economic prosperity," wrote a columnist for the Madison Capital Times.

Carl Zichella of the Sierra Club characterized the legacy of sulfide mineral mining as among the saddest of any industry. "Now Exxon - whose track record, it must be said, is as soiled as the shores of Prince William Sound following the catastrophic oil spill they caused there - would have us believe that things will be different here in Wisconsin." The Exxon Valdez "was on everybody's radar screen," said George Rock, who printed a bumper sticker, "Save the Wolf River from an Exxon Disaster," displaying a ship leaking oil. A Green Bay journalist, Mitch Bent, referred to the events surrounding the Exxon Valdez when he commented, "We're told we should 'trust the process,' that we should 'trust the DNR,' that we should 'trust the mining companies.' Why should we?"[16]

Consultants hired by the DNR and by anti-mining groups refuted some of the mining company's rosy projections. In a report to the DNR, Bill Freudenberg, a professor in the rural sociology department at the University of Wisconsin-Madison, disagreed with the economic picture drawn by the Crandon Mining Company. The company claimed the mine would bring hundreds of long-term mining and mine-related jobs to Forest, Oneida and Langlade counties. Local communities would receive about $72 million in tax payments and $43 million in spending for goods and services during construction. During the mine's twenty-five years of operation, the spending of workers would stimulate the local economy.

Freudenberg collected over 200 studies of mining towns and found that most of them showed few long-term economic benefits from mines. He accused the company of not considering the volatility of the worldwide market for zinc and copper. Using zinc prices as an example, he pointed out that the mineral's price was 30 cents per pound in January of 1986, 90 cents per pound in January 1989, and 50 cents per pound in the middle of 1991. "Volatility was more the rule than the exception."

Consequently, Freudenberg said, the operation of the mine was likely to fluctuate depending on the price of zinc and copper. When the market was good, the mine would operate, but when the market was down, the mine would close or slow down, and the miners would be out of work. "If I were in their shoes - that is the people of Crandon - I wouldn't believe the numbers coming from Crandon Mining Co.'s predictions," Freudenberg concluded.[17]

Another consultant, Charles Norris, hired by the Sierra Club, debunked the company's EIS assessment of virtually zero impact on the environment. "Hundreds of people will descend upon the woods and wetlands near Rice and Mole lakes," said Norris, "undertake major engineering and construction; excavate, digest, transport and/or dispose of 55 million tons of bedrock;...export billions of gallons of water; and, after the better part of three decades, dismantle the facilities and leave. At the height of this activity, the EIS predicts negligible, almost immeasurable, impacts....It is an assessment that is intuitively too optimistic. And, in this case, intuition is correct." The company's EIS, said Norris, was a "promotional vehicle that shuns critical investigation, obfuscates the assessment of data that is available, and stretches credulity in its conclusions."

Norris described several impacts that would likely be greater than the Crandon Mining Company would admit. More groundwater would move through the ore and the area than acknowledged, and more groundwater would have to be pumped; more acid mine drainage than predicted would occur; more water than predicted would contact the abandoned mine and waste disposal areas after reclamation; finally, groundwater and surface water would be "permanently adversely" impacted by the project.[18]

Douglas Cherkauer, professor of geo-science at the University of Wisconsin-Milwaukee, who criticized computer models during Exxon's first go-around, repeated his critique of the new project. CMC's categorization of two lakes, Mole and Rice Lake, "leaves me baffled." The company claimed that the lakes would not be impacted based solely on the groundwater flow model, but the computer model used to set the lake levels was defined <u>a priori</u>. "Once defined,...it can't change." Then he concluded:

This strikes me as circular reasoning. The mine will have no impact on the lake levels because the model, which is programmed to not allow changes in lake level, shows that lake levels don't change. Unless I'm missing something here, I think it has to be admitted that the presently-configured impact assessment process cannot define impacts to Mole and Rice Lakes. The WDNR and USCOE should require complete reconfiguration of the ground-water models on the west boundary. Without that, the Sokaogon Community should have no confidence that the impact assessment process and any mitigation plan stemming from it will address their concerns.[19]

Critics of the company's plans thought that modeling for the aquifer system in Forest County was impossible to replicate and predict. With the area's irregular rainfall, complex glaciation, and many lakes, streams, and rivers running in different directions, the company could not have chosen a worse place to mine. "It's probably the most complicated aquifer system on the plant, and this is undoubtedly the most complicated model ever attempted," said Phillip Seem, the consulting environmental engineer for the Menominee.[20]

*　　*　　*

Never was there a coalition opposing the Crandon mine project like the one that formed in the spring of 1994. The coalition was inclusive rather than exclusive. Indian tribes, fishing and hunting organizations, and conservation activists, fifty diverse groups, succeeded in coming together in an aggressive campaign against the mine The diversity of the movement provided a hedge against attempts to marginalize mine opponents as a radical fringe. "We had the rich to the poorest of the poor; it was a collection of everybody," said Menominee leader, Ken Fish.[21]

"Keep fightin' 'em, son!" "Keep fightin' 'em!" lake property owners told Sonny Wreczycki. His group of volunteers held fundraisers, wrote letters to newspaper editors, attended hearings, and collected thousands of signatures on petitions. In a televised debate with the company's Ken Collison in Wausau, Wreczycki lashed out at his opponent. "You don't know a damn thing! You don't know what impacts are going to be here!"

Every Labor Day weekend, the Rolling Stone Lake Association sponsored a picnic. On that weekend's Sunday, a large crowd gathered to hear Wreczycki make his annual updated mine presentation.[22] One of his new efforts was to disprove the company's contention that Little Sand Lake was perched, meaning that the lake was like a bowl with no water entering or leaving. "I knew better than that," Sonny recalled. "I swam bare-ass there in the early 1960s. I could feel those little cold springs running up [my] butt."

Sonny recruited a town supervisor's son, a scuba diver in the Appleton area, to dive with a camera into Little Sand Lake and document the springs. An initial dive in September 1995, located a likely spring in the northeast end of the lake. A month later the divers relocated the original spring, gathered samples to document the discovery, and found four additional springs. Videotape shot during the underwater exploration showed approximately 5,000 square feet of lake bottom where sediments were absent. Wreczycki had rock samples from the spring areas analyzed by the United States Geological Survey. The rock samples indicated groundwater inflow to the lake. Wreczycki showed his video at a DNR meeting to prove his point.

"The springs were not described by CMC prior to their discovery and have therefore not been included in estimates of the lake bed conductivity values now in use in the regional groundwater modeling," a Wreczycki ally wrote the DNR's Secretary George Meyer. Meyer should see to it that the DNR staff study all groundwater inflow sources in Little Sand Lake. "The springs' ability to transmit back-filled waste rock contaminants from the ore body itself, as well as from the Tailings Management Area (TMA), should be quantified and subsequent impacts to Little Sand Lake and the downstream watershed estimated." Subsequently the DNR drilled through the ice on Little Sand Lake, found the springs, and modified its computer modeling.[23]

Wreczycki's health suffered during his anti-mining activity. He endured rheumatoid arthritis and experimented with powerful drugs to relieve his pain. Flare-ups occurred when he was under stress, often during meetings and public hearings. The drugs weakened his immunity, and he would pick up some virus or another. "He would spend [four days to a] week recovering from whatever bug he picked up," his friend Mike Monte observed. "All it took was for someone to cough and Sonny would be lying on the couch." In spite of his wavering health, Wreczycki persevered and attended scores of meetings. "It was a huge commitment," said Monte.[24]

Sister Toni Harris resumed her attempts to inform Exxon shareholders. At Exxon's annual meeting in Fort Worth, Texas, her resolution secured 5.3% of the vote. Speaking on behalf of the Sokaogon, Fred Ackley reminded shareholders that his tribe was "fighting for its life." Lee Raymond, Chairman of the Board and CEO of Exxon, criticized the tribe for not responding to Exxon's offer to discuss their mutual concerns. Raymond quoted from Ackley's statement in the New York Times that "Talking with them is participating in our own destruction." Responding to Raymond, Sister Harris cited the experiences of many Indian tribes who had sat down with corporate executives, expressed their concerns, only to have them ignored in corporate decision-making. "Then the corporation turns around and claims that the Indians were participants in the decision-making process."[25]

Several new groups, not present during the 1980s, now supported the anti-mining movement. The Midwest Treaty Network, an alliance of Indian and non-Indian groups that supported the treaty rights of Native Americans, had been founded in Lac du Flambeau in 1989 during the Chippewa spear fishing struggle. It focused on defending and strengthening Native cultures, "protecting Mother Earth and fighting racism." At boat landings during the tumultuous spear fishing seasons in Wisconsin, the network stationed hundreds of Witnesses for Nonviolence to document and divert anti-Indian harassment and violence. After the spear fishing controversy died down in 1991, the network turned its attention to asserting tribal sovereignty and treaty rights to protect northern Wisconsin waterways from the Crandon mine. The Network's publications stressed "using basic English and easy-to-understand visuals." It developed 60,000 informational brochures on the mine and distributed them through the tribes, sporting clubs, and rural shopper newspapers.

The Midwest Treaty Network's primary leader, Zoltan Grossman, worked in Madison as a cartographer at a private firm. In 1996 he entered graduate school at UW-Madison, received his Ph.D. in 2002 in Geography and American Indian Studies, and the same year began teaching at UW-Eau Claire. Grossman won the lasting gratitude of the tribes. Open-minded, an exceptional listener, he "felt what we were feeling inside about what the mine would do to our culture," Tina Van Zile observed.[26]

In 1995 Grossman organized the Wolf Watershed Educational Project (WWEP), focusing on the Wolf-Fox River waterways. The project aimed to educate local communities about metallic sulfide mining, cultivate local community organizing skills, and establish links between Indian and non-Indian communities. The project assumed apathy had changed the nature of organizing. "We cannot build understanding by expecting the public to come to our conferences or to government hearings," said a press release. "Instead, the project stresses group members going to people in their own communities, their own clubs, and at their own events. The key is visibility - and strategy for environmental justice will be rendered ineffective if organizers only talk with each other and government officials, and don't reach thousands of people that have not been reached before."

The project did reach thousands of others. Two speaking tours were scheduled in twenty communities near the Wolf River and the Fox River. Each tour made a conscious effort to include speakers from environmental, sport fishing, and Native American groups, to prove that mining opponents could work together. "For a lot of people who came, it was the first time in their lives that they sat down in an audience and heard a Native American speak," said Grossman. Among the speakers were Herb Buettner, George Rock, and Ken Fish.[27] Each event was covered in the local newspaper that often featured text of the speeches as front-page news. Afterwards local activists would continue to incite interest and generate phone calls and letters.

On May 4, 1996, the tour drew 1,000 people to a rally at the proposed wastewater discharge site on the Wisconsin River. In fall 1997, the WWEP sponsored a Circle Speaking Tour around the edge of the state, reaching twenty communities where mining was not yet an issue. A fourth tour later focused on schools and universities. Company officials were dismayed but impressed. "Brilliant, masterful," said a company consultant of WWEP's efforts.[28]

The Mining Impact Coalition in Madison also "spread the word." Its founder, Dave Blouin, an active Sierra Club member, and volunteers who assisted him were experts in sending out press releases, faxes and e-mails. Because northern Wisconsin activists had difficulty traveling to Madison, Blouin monitored the legislature, videotaped many technical meetings between the company and the DNR, and studied, copied and distributed the open records at DNR headquarters in Madison.[29]

Activists also formed several local anti-mining organizations. The most active emerged in Shawano County. For years Len Pubanz, a biology instructor at Shawano High School, had fished on the Wolf River almost every day after he came home from school. For thirty-five years Len and his wife, Judy, lived on seven acres of wooded land on the Wolf River and appreciated its "picture-postcard aesthetics." Judy, an artist with a studio at home, enjoyed painting scenes on the river. "We knew nothing about the proposed Crandon Mine," said Len. But in 1996, after the couple attended a session of the Wolf Watershed Educational Project, they became fervent anti-mining activists. "We had never before been involved in anything of this sort - demonstrating," Judy recalled after a rally in Rhinelander. "We felt like hippies."

On June 11, 1996, they arranged an organizational meeting and formed POW'R (Protect Our Wolf River). Over the next several years, volunteers attended their monthly meetings. "We're going to make people aware, down to Clintonville and New London and to the south," said Judy. In the Town of Wescott, they brought resolutions to town officials asking them to take a stand against the mine. The first resolution passed and so did many others presented to Shawano County municipalities by POW'R. For the next eight years, the anti-mining movement "engulfed" the couple's lives.[30]

Fishing groups were also crucial components of the broad coalition. Trout Unlimited, Walleyes for Tomorrow, and Sturgeons for Tomorrow opposed the mine and passed anti-mining resolutions. They were important because public opinion and many legislators viewed them as conservative, non-partisan "average folk."

Keep the Wolf River pristine, argued the fishing organizations. "Our members have fished throughout the American West and have seen the tragic after effects of mining on trout streams," said John Welter of Trout Unlimited. "We don't want to see those things happen here."

"Trout have to have clean water to live," said another member of Trout Unlimited. "Trout are like the canary in the mine shaft; if the trout die off, the water is too dirty to use. The trout is too valuable a resource to just waste."

Walleyes for Tomorrow had twelve chapters in Wisconsin with 3500 members. Tom Soles, the state president, said, "The Wolf River is the key to the Lake Winnebago Walleye population. If the sulfur ever came down the river that far, Winnebago would stop as a hatching facility."[31]

Herb Buettner, along with George Rock and other non-Indian opponents, helped link the Menominee with angling groups. Rock told the sport fishers that he "trust[ed] the Menominees with walleyes more than [Governor] Thompson." Tom Soles contacted Ken Fish, and the former adversaries in the spear fishing conflict agreed "we can sit down as friends." Fish worked with Soles and other sport fishing leaders on a weekly basis in support of the anti-mining cause. "It was surprising to see people who were yelling and screaming at tribal members on the [boat] landings, later testifying together against the mine proposal at the capital," recalled journalist Ron Seely.[32]

124

Anti-mining rallies were usually well attended. On March 14, 1994, three hundred people, many from northern Wisconsin, gathered at the State Capitol in Madison. Articles describing the rally made the front pages of the Milwaukee Sentinel in Milwaukee and the Wisconsin State Journal in Madison. "Marching to the chant of Indian singers and the beat of three large drums, the crowd carried its protest to the doors of the DNR and Wisconsin Manufacturers and Commerce," a reporter wrote. "Just in case Exxon executives in Houston may have missed it, I'm sending them a press packet with promises of more to come," Al Gedicks wrote in his newsletter.

George Rock   Herb Buettner

Public hearings were often tumultuous. "Liars. Traitors. Pawns." Those were some of the terms fired at DNR staff members during a raucous hearing at the Ainsworth Town Hall in Forest County. Bill Tans, Chris Carlson, and Archie Wilson, the DNR's representatives, found themselves on the defensive before an ornery crowd. Tom Ward, 67, who operated a summer camp for boys with disabilities in Forest County, was the most boisterous. "They're cheerleaders for the mining company!" Ward said of the DNR. "They're not a citizens group - they're a political arm of the governor. That's what makes people so mad up here....The governor gets rid of the [state] public intervener's office and then makes the DNR a Cabinet agency so he can control its every move. And he thinks people can't see through that? It's outrageous. The DNR can't be trusted. It's that simple."[33]

John Mutter Jr., active for two years in the Shawano-based POW'R organization, wrote about his experience in his book, To Slay a Giant. A power plant worker and freelance writer from a small town in Shawano County, he regarded his experience as a "hands-on civics lesson." He had never been involved with an environmental group before. "I didn't know much about politics and what it took to get a bill passed."

"When one gets involved in something like this, you just have no idea the degree of intensity, or the length of time that's going to be expended," he said. "This issue was like swimming near a whirlpool; suddenly I was pulled in and deeply committed." The mining company had money, but "we had people, and people in the long run are going to win."[34]

Mike Monte, publisher of the Pioneer Express, was an outspoken mine opponent in the Crandon area. A former logger, he had supported the proposed mine in the mid-1980s, but now changed his mind. "A decade ago, I thought the mine would have brought jobs here," Monte said. "Now, I believe you would give up more than you get from this mine. Our county is water-rich. It's not a proper place to mine....A mine is not my idea of what northern Wisconsin is or why tourists come here."[35]

Mine critics seemed almost obsessed with the mine controversy. Most didn't care if a colleague was a Democrat or a Republican. "[Partisan] politics never entered into it because that was not our focus," said Maureen (Mimi) Wreczycki. She couldn't remember ever getting into a partisan political discussion with anybody. "It was always the mine."

"Everybody can imagine a football field," said one opponent. The Crandon mine's tailings ponds equaled "365 football fields." Al Gedicks used another metaphor. The mine would generate about 44 million tons of wastes -"the weight of eight Great Pyramids."[36]

"We didn't have a place to hold meetings," said Sonny Wreczycki. "So we had them in homes." George Rock hosted many "kitchen table" meetings where his emotional state would fluctuate.

126

If the cause had lost a recent court decision, Rock was depressed, but he'd recharge at a kitchen table meeting. "We never had a lot of fundraisers," recalled Rock. When his anti-mining associates needed stamps and food, they reached for their wallets. (Rock spent thousands of dollars of his own money.)

Kitchen Table Meeting

Rock's zeal bordered on paranoia. He worried about the powerful mine owners. At a demonstration in Ladysmith, he noted that police had written down the vehicle license numbers of opponents. He even carried a shotgun to a meeting in Nashville. "Maybe the mine was risky, but isn't there always risk in life?" Rock was asked. "Yes," Rock responded, "but who's profiting? What's my profit? They are going to make money for stockholders and are going to be gone, and I'm going to be a loser, loser, loser. It's my environment that we're risking."[37]

Zoltan Grossman tried to arrange for Rock to be interviewed on a major Canadian radio program, listened to "by millions." "It's these two guys who do 5-minute taped interviews," wrote Grossman who had faxed the station anti-mining literature. "If you are interviewed keep it short and succinct, but get the main points in," he advised. "Spell out Rio Algom, and ask Canadians to call their office in Toronto....This is a really important interview - mention that we want Rio Algom to withdraw...for the honor of Canada."[38]

"There is no silver bullet," Carl Zichella advised mine opponents. "This is going to be death by a thousand cuts." He might have said a hundred thousand cuts if he had counted the anti-mining messages generated. To stop the proposed mine, people were repeatedly advised to write, e-mail, fax or phone the U.S. Army Corps of Engineers, Exxon Chairman Lee Raymond, Wisconsin Governor Tommy Thompson, the DNR's Bill Tans and DNR Secretary George Meyer. Addresses and phone numbers were provided.

"If you're from Wisconsin," opponents urged people to form a local group, help with time or money, pass a resolution in their community, write letters to the editor, call radio talk shows, inform their family, friends, and community leaders, contact State Senators and Assembly Representatives, use a toll free hotline for information about the proposal (1-800-432-8747), attend local, state, and federal hearings, and support tribal efforts to strengthen their environmental enforcement under the Federal Clean Air and Clean Water Acts.

"If you're outside Wisconsin," people should call the Wisconsin Tourism Bureau, ask about the impact of mining plans on tourist spots, "help us make the Crandon mine a national environmental justice issue," place an article in the newspaper or a story on radio/TV, attend a rally, or carry a sign. "Get your local county or town board, sports, senior and community groups to pass a resolution opposing metallic sulfide mining. Pass a resolution in your own community group."[39]

By January 1998, forty-two counties, cities, villages, and towns along the Wisconsin River had passed resolutions against the Crandon Mine, the wastewater pipeline or both. Fifty-two other counties, villages, towns, unions and organizations had done the same.

In an attempt to educate Illinois residents about the mine, the Wolf Watershed Educational Project provided speakers to talk about the mine at a rally in front of the Wisconsin Travel Information Center on North Michigan Avenue in Chicago. "A lot of Illinois residents have visited the Wolf River in recent years and have expressed concern about the proposed Exxon mine," Zoltan Grossman explained.[40]

In July 1997, Earth First!, one of the most militant environmental groups in the world, came to Crandon to protest the mine. In the past at rallies around the world, Earth First! members had chained themselves to bulldozers, climbed trees slated to be clear-cut, built roadblocks on logging roads, sabotaged power lines and vandalized machinery. The group believed mainstream environmentalism had been a failure.

In Crandon the mining company responded with locked doors and security guards. With Earth First! nearby, some company employees wouldn't open their mail at the office or at home. Rumors spread that people in the forest industry had received bombs in the mail. "I'm sixty years old and thinking of retirement," recalled Dick Diotti, a company spokesman. "I like to play golf. All I need to do is be missing four fingers."

At their rally eighty, protesters of Earth first! waved signs, chanted and pounded on the locked door of the Crandon Mining office. "Some of the protesters spread an Earth First! banner with the organization's green-fisted symbol in front of the office. Others laid on the sidewalk in front of the office." Twenty-nine were arrested. Wisconsin opponents of the mine stayed away from the rally, reluctant to be tagged with a radical label. One observed:

We are very careful not to connect ourselves with these people,.....We work with legislators on a political...playing field, and we can't appear to be those kind of people, that do something illegal, or against the public interest....If the [mining company] could tie us to Earth First! or some radical outfit, and they try, they would do it in a heartbeat.[41]

On Thanksgiving Day 1995, Alice McCombs of Shawano started online anti-mine activism with EarthWINS Daily, an email newsletter she distributed to 500 people and organizations around the world. Several months later, while working as a research analyst for the Menominee Treaty Rights and Mining Impacts Office, McCombs designed the Menominee website about the Crandon mine, the first website focused on the controversy. McCombs recalled:

This was back in the days when the general public was just starting to use email and the World Wide Web was in its infancy. The Wisconsin legislature and Department of Natural Resources were amazed to see emails coming in regularly from Wisconsin citizens, other states and other countries who were strongly opposed to the Crandon mine.[42]

By 1997 a dozen Web sites had been launched by groups opposing the mine. The sites kept people informed of meetings, rallies and hearings. "The Wisconsin Stewardship Network's page on the Internet included a feature that allowed visitors to click on a map of the state and call up the names of various environmental groups and leaders in whatever part of the state they were interested in." Internet sites allowed activists in Madison and Milwaukee to work closely with northern Wisconsin allies to improve grassroots efforts.[43]

Critics of the mine also used other electronic communications to lobby politicians. A mine opponent explained:

I call [my contact] on the phone and tell him I will fax him a press release. I draft the press release on the computer. I fax it to him on the computer. At times, I will fax to him by 11 a.m., and he will do the release, and by 4:30 p.m. that is picked up by newspapers across the state. News has to be right now....We can get right to those reporters and publishers, right now.

When using Google, the popular computerized search engine to access information about the mine, the first response read "No Crandon Mine." A mining company executive sadly observed:

If you were to get on the Internet and type "Crandon Mine," you do not get our web site. You get all the other ones. If you go into any one of the groups, they are all linked. They are very good at using the Internet as a tool. And if I'm a person who just wants information, I get all of theirs first. Whatever the key phrases they use they are very good at it.

The journal of the National Mining Association bitterly complained that Wisconsin's "barbarians in cyberspace" were distributing anti-corporate messages throughout the world.[44]

Company officials thought the media held mine opponents to a lower standard. "Whatever we said or did had to withstand the test of public scrutiny....What we said had to stand up," said Rodney Harrill. "Our opponents didn't have the same standard....They could say whatever they wanted - fact or not fact....There was no accountability. They didn't have to prove what they said was right."

Company engineer Gordon Reid held a similar view. "It was frustrating to me that, after we'd spent a million dollars on a study, with the foremost experts in the country on the stand during a hearing, that we'd have Joe Blow, a retired used car salesman from Sheboygan, [and] his word was taken with as much weight or more [than our experts]."[45]

A degree of hypocrisy was imbedded in the anti-mining movement. Many opponents wanted mining stopped everywhere, yet they drove to protest rallies in autos (built from mining products), powered by gasoline (refined from mined petroleum). "I saw no horses tethered to posts outside the hall that night," said Mitch Bent of the Green Bay News-Chronicle. "Making light of lower-paying jobs and a 25-year lifespan for the mine won't win environmentalists any friends, either. Driving a truck for $8-$12 an hour looks pretty good to someone earning minimum wage. And twenty-five years is a career for most people."

Opponents were sometimes crude and they often exaggerated. "I am requesting an answer to a rather gruesome question," a critic from Bowler, Wisconsin, wrote to Bill Tans. "What is your best estimate of the body count at the Crandon Mine? How many body bags will be necessary? The corpse count?"[46]

When CMC "comes to town and states that its mining intentions are 'strictly honorable,' they have the same credibility of a convicted pedophile who applies for work at a day care center and claims that he or she is cured," wrote an anti-mining columnist.

The DNR was often criticized for using exaggeration and distortion. The department's "pro-mining bias," said opponents, was "pervasive" and "blatant." It was driven "exclusively" by "political considerations."

Al Gedicks charged that besides sulfuric acid, the mine could produce "high levels of poisonous heavy metals like mercury, lead, zinc, arsenic, copper and cadmium, when exposed to air and water." Gedicks grossly exaggerated.

Gedicks also contended that during negotiations for the local agreement between the Town of Nashville and Crandon Mining, the Mole Lake Sokaogon "was not even consulted." In truth, the Sokaogon were repeatedly approached by the company, but they chose not to consult. In Gedicks' view and in the view of many of his associates, company officials and engineers were not professionals deeply committed to their work, never had high motives, and were always conniving.[47]

Gedicks' speeches, newsletters and newspaper commentaries enraged company officials. "Gedicks would put out one of his documents, with all the footnotes, and [company] people would pick it apart internally and point out all the misrepresentations and all the misstatements," noted a company consultant. "He would just drive everyone nuts....But he was very effective."

While some mining opponents often exaggerated and distorted, some didn't and were well informed, even on highly technical issues. "I had not anticipated this level of knowledge, this far from the mine site," said Archie Wilson, DNR's field coordinator for the mine project, after a meeting in New London. "It surprised me."[48]

*   *   *

Since July 1994, the company had been trying to negotiate local agreements with several jurisdictions. "Although we believe we are close with Forest County and Nashville, we have not as yet completed any agreements," Jerry Goodrich wrote to his superiors about the local pacts.

He explained the process and the need for the pacts. "Under Wisconsin law, local agreements allow local zoning jurisdictions to waive their zoning and permit ordinances...and through the local agreement contract, grant a mining company all the permits it needs to operate in their zoning jurisdiction....In exchange for signing a local agreement, the zoning jurisdiction can extract whatever it can negotiate from the mining company in terms of money and other commitments."

Although the agreements were voluntary and not required by law, the alternative was to apply for each permit required from each jurisdiction and pay the fees attached to them. "If we attempted to go the permit route, we believe local jurisdictions would very quickly pass anti-mining ordinances."

Goodrich hoped to get at least one jurisdiction to sign an agreement soon. If one signed, others would probably follow. "We must be able to persuade legislators [in Madison] that the people most directly affected by the project support it." Given the length of the time the company expected to be operating in Crandon and its huge investment, it made sense "to try to stabilize as many pieces of the regulatory environment as possible, as early as possible. The local agreements will help us do that."[49]

The local agreement arranged with the Town of Nashville called for the Crandon Mining Company to pay the town a $100,000 permit fee and annual payments of $120,000 for up to six years. The company also guaranteed several additional payments, including up to $500,000 in legal expenses, $150,000 to finance a citizen advisory committee during the life of the mine, and $60,000 for repairs to Little Sand Lake Road. (The Town of Lincoln signed a similar agreement several months after Nashville.) Forest County would receive a total of $1.9 million in guaranteed payments in the first seven years of construction and operation. "I'm sure our taxes will be dropping if this mine is approved," said Nashville's board chairman, Richard Pitts, a lifelong resident of the town.

In exchange for the guaranteed payments, the three units of government agreed to support the company's application for mining permits and specifically exempted the company from all zoning ordinances, regulations, and laws.

(In a minor agreement signed with Crandon, the city approved a zoning variance for a 165-foot-tall mine head frame. No building was to be taller than fifty feet within several miles of the Crandon municipal airport.)[50]

Critics contended the agreements were premature, coming before the DNR completed its investigation. "What's the hurry?" asked Neil Schallock, a member of Nashville's zoning committee that had been excluded from the closed negotiations. It did not guarantee jobs for any local residents, only "preference" and "to the extent permitted by law" for applicants who had lived in the area for "one year."

Limitations were placed on reopening the agreement. Unless residents suffered additional "unmitigated negative impact" or the operation was "substantially different" from the mining plan, the mining company could prevent the town from reopening the agreement. Regarding catastrophic chemical accidents and other disasters, the company simply said it had a plan and would maintain equipment and personnel for accidents that occurred at the mining site.

At a meeting to discuss the proposed agreement, Kevin Lyons of Milwaukee, Nashville's attorney, tried to avoid hostile questioning and stated town officials could ask questions after CMC President Jerry Goodrich and Don Moe gave their presentations, but the public in attendance could ask questions of CMC officials in private only. David Anderson, a Nashville town resident, rose and objected. "This is undemocratic for an out-of-town lawyer to prevent people from asking questions of CMC officials in the presence of elected officials. As a resident of this township, this is my democratic right."

George Rock spoke. "There has been a lot of discussion of the relative risks of this project," said Rock, "Exxon/Rio Algom is proposing to construct the largest toxic waste dump in the state here, but they cannot provide a single example of where this has been done without harm to the environment. This is like an intersection where no car has ever made it across without being hit by another car. Do you want to allow your child to cross that intersection?"

"I've never heard such a one-sided presentation," complained Tom Ward, president of the Crandon chapter of the Wisconsin Resources Protection Council, as he addressed town officials. "We're getting a bad deal here. They [CMC] get all the profit and we end up with all the metallic sulfide waste. Their models tell us that we won't be affected by acid mine drainage and heavy metal pollution. But what if their models are wrong?"[51]

The leader of the opposition was Chuck Sleeter, a former sheriff's deputy from Wood County, who three years earlier had fulfilled a lifelong ambition to build a home on Pickerel Lake, then retire there to hunt and fish. "I want CMC to guarantee that my water, air and lifestyle will remain unchanged. Can you give me that guarantee?" Sleeter demanded. Don Moe, CMC's technical manager, replied that he could not.

Joanne Sleeter   Chuck Sleeter

When the agreement was finalized and made available to the public, there was outrage. "My phone hasn't stopped ringing," said Sleeter. "I've had people call me half crying over the agreement. We are going to take the steps necessary to stop this agreement from being signed." Over two hundred Town of Nashville residents signed a petition for a special town meeting on Saturday, December 7, 1996, to approve a resolution preventing the signing of the local agreement. The town board scheduled a public hearing on Thursday, December 12, at the Nashville Town Hall.[52]

In a letter to Richard Pitts, attorney Lyons advised against accepting the "petition" and "resolution" at the special town meeting. The town meeting could not *direct* the actions of the town board; it could only authorize certain actions of the town board. "The meeting has no authority or jurisdiction over the local agreement. The resolution as proposed is null and void." As chair of the meeting, Pitts "should declare the motion out of order and prohibit a motion for [a] vote. Even without such prohibition, a vote would result in an action that was null and void and would have no legal effect."

On the evening of December 7, opponents of the mine submitted 235 signatures on a resolution to vote down the local agreement. As a large crowd entered the town hall, Lyons advised Pitts on his options. The chair could conduct a vote or adjourn the meeting. Unaware that a radio microphone was turned on, Pitts asked Lyons what he should do. Lyons responded, "It's your choice; you can let people talk, you can take a vote, or you can just adjourn it. It's really your call." When Pitts decided to adjourn, Lyons warned, "If you do what you [are] planning on doing, then you...better stand up and walk, so people don't think they can continue yakkin' at ya."

Pitts opened the session, then quickly adjourned. The crowd booed loudly. Critics yelled, "Look at all the people who are here, can't you listen to us?" "This is America? This is democracy?" "Pitts is a dictator!" Someone called the police to maintain order.

At a follow-up meeting on December 12, over 200 people crowded into the town hall for the public hearing. Tempers immediately flared. The first speaker, Tom Ward, intoned, "Lincoln said 'government of the people, by the people, for the people.'" Instead, "What we're seeing is government of Exxon, by Exxon, for Exxon. This agreement is a disgrace. It's an absolute outrage. It's a travesty of the democratic process."[53]

A Chicago Tribune reporter described the end of the emotional three-hour meeting. Dozens of town residents pleaded with their board to reject the local agreement. As the stone-faced, three-member board voted to approve it, audience members screamed at them, calling them "traitors" and "dirty, rotten scum." "Shame on you for selling your neighbors out," an older woman yelled as the meeting drew to a close.

When the audience asked board members to explain their vote, they refused and instead asked attorney Lyons to outline the agreement. The crowd didn't want Lyons. "We didn't elect you!"

someone shouted at him. "You can go back to Milwaukee. What do you care?"[54]

Opponents were also upset that the town boards of Lincoln, Nashville, and Forest Counties scheduled these critical hearings at the same time on the same date, December 12. It appeared to be a blatant attempt to dilute the numbers of attendees and comprehensive media coverage. "Why do we have three meetings the same night?" asked Neil Schallock. "I guess it's obvious why we do."

That evening in Crandon, the Forest County Board approved its own local agreement with the mining company, 18-3, even though testimony at the hearing ran four to one against the agreement. "They are tampering with our most precious resource, our water," said the author of a letter read into the record. "When it is over, they'll leave us standing here, holding an empty bag."

When Attorney Kevin Lyons negotiated with the Crandon Mining Company, he not only worked on terms of the local agreement, but he also negotiated for his legal fees to be paid through the town by the company. It was deemed likely the company had no intention of paying Lyons if the town didn't approve the pact, and so he had a vested interest in the approval of the local agreement and a potential conflict of interest when representing the town in the negotiations.[55]

Thereafter opponents of the local agreement remained bitter at the outcome and often bemoaned the squashing of democracy. "What happened to democracy?" asked Jim Wise. "Democracy is like a muscle, we must use it or we will lose it."

Normally town board elections receive little attention in Wisconsin, but the Nashville election on April 1, 1997 drew considerable interest. Chuck Sleeter and a slate of mine opponents challenged the incumbents who signed the local agreement. John Nichols of Madison's Capital Times noted that the "Nashville election will do precisely what politics should do: It will focus attention on a vital issue of statewide concern, and it will give ordinary citizens a rare opportunity to challenge a corporate initiative that is almost certainly not in their best interest....This year, that eternal struggle for a real and meaningful democracy begins in Nashville."

On the day of the Nashville election, George Rock trucked voters to the polls. "We made sure that everybody in Mole Lake voted," Rock recalled. "We wanted everybody who wasn't ambient ground temperature." Sleeter and his compatriots won a stunning victory.

The anti-mining victory in Nashville altered the political power of town governance as it dealt with the mine issue. Now it was not just Wreczycki's Rolling Stone Lake group and the Sokaogon Chippewa who opposed the mine. "Now we had a town government inside of the mine [area] against the mine," Wreczycki noted. Subsequently Sleeter survived a recall election and a regularly scheduled election with his anti-mining board intact.[56]

Almost immediately following his election victory, Sleeter and the board rescinded the local agreement. The company sued the town, and after several court decisions and appeals, the company won. The local agreement was declared legal. However in the final legal stipulations, members of the old board had to concede that they had participated in twelve closed meetings between November 1993 and December 1996. None of the notices for the closed meetings were sent to the local media as required by Wisconsin law; nor did posted notices of closed meetings indicate the subject of the meetings as required.[57]

Public opinion polls increasingly favored the anti-mining cause. In April 1997, a public opinion survey conducted by Wisconsin Public Radio and St. Norbert College indicated that 48% of people surveyed opposed the development of the mine. 31% favored it. Opposition was strongest among liberals (54%), those aged 18-34 (62%), those who thought the anti-pollution laws in Wisconsin were not strict enough (56%), and among those who judged that Wisconsin's environment had worsened in the past five years (68%).[58]

The DNR's Chris Carlson thought opposition to the mine and related criticism forced the DNR "to do the right thing." "It provided us with better ammunition with which to address the applicant in getting a proper record...for a decision." Without opponents, "I am sure the process would have been quicker," and "the level of public scrutiny...would not have been as extensive."

"It was the most effective group of environmentalists and others that I've ever seen across the state," said Harold "Bud" Jordahl, a long-time environmental leader in Wisconsin. "Usually these groups are highly territorial. They stake out their own issue and they don't like to look beyond it. It's a very difficult thing to pull these groups together."

"This is not just a ragtag group of individuals," observed company president Rodney Harrill. "It's an organized, coherent opposition. At different meetings I see them reading off the same script that's printed off the Internet....Our opposition was way out ahead of us on this thing."[59]

CHAPTER SEVEN

MORATORIUM

Back in the 1980s, Evelyn Churchill, Roscoe's wife, had suggested a statewide mining moratorium, but the idea floundered for over a decade. In December 1994, more than 7,000 people signed petitions appealing to the state's Natural Resources Board to support a mining moratorium, but the board ignored the appeal.

In late 1995, over lunch in Madison, opposition leaders launched an aggressive statewide moratorium movement. Democrat State Representative Spencer Black recalled "The message was that the mining companies did not have the technology to undertake a sulfide mine without polluting the environment, and Wisconsin should not be a guinea pig for unproven technology."[1]

Supporters of the moratorium argued that engineering projections were not the same as real science that demanded proven results from past experience. Since no one had proven that a sulfide metallic mine had ever been successfully operated and reclaimed without environmental degradation, Wisconsin should demand such an example from the mining companies before they experiment with the state's precious resources.

Initially, prospects for passing a moratorium appeared slim. It wasn't necessary, said proponents of the mine, including the governor and the Republican controlled legislature, because strict laws on mining were already in place. "Capital pundits said it didn't have a chance," said Black.[2]

In early April 1996, Crandon Mining Company sponsored full-page ads in major state newspapers, trying to sidetrack any momentum for a moratorium. "Those who want you to oppose mining have the same opportunity to offer credible proof - evidence - supporting their claims," the ads said. "We want you to form an opinion too. We just hope it will be based on facts and science."

Critics of the moratorium thought it was flawed conceptually because it singled out one industry: mining; the proposal was merely a "politically contrived, simplistic effort to circumvent the regulatory process of the state Department of Natural Resources."

The moratorium bill had no support and was only a gimmick, several Republican legislators declared. Republican Assembly Majority Leader Scott Jensen called Black's efforts on behalf of the bill "pure demagoguery." Wisconsin "should not turn its back on developing technology."[3]

The moratorium wasn't necessary, George Meyer asserted, because the DNR could effectively regulate the mine. The decision by Meyer to insert himself into the debate as an advocate against the moratorium "was one of the most serious breeches of public duty in the 11-year history of Gov. Tommy Thompson's administration of the state," the Capital Times editorialized. "By siding with Exxon...Meyer has leant credence to the argument that the Department of Natural Resources is rapidly becoming a politicized arm of the Thompson administration."[4]

Despite opposition to the moratorium, grassroots organizing propelled the proposal forward. In May 1996, an avalanche of calls and letters by moratorium supporters captured the attention of legislators and forced them to think seriously about the measure. In the first two weeks of the month State Senator Robert Cowles, Republican from Green Bay, received over 350 contacts on the mining issue, almost all of them from opponents of the mine. "I was hearing it everywhere I went," said Cowles. "There's a lot of pressure building," agreed Republican State Representative DuWayne Johnsrud, chairman of the Assembly's Natural Resources Committee. The Crandon mine issue was a "red hot topic in this area," a Shawano resident wrote Bill Tans.[5]

In a report to Don Cumming in July, Dale Alberts expressed dismay at the eroding support for the mine project and worried that a mining moratorium bill would soon pass in the legislature. "The Crandon Project is in serious and immediate jeopardy." The company's polling data showed a dismal trend. Most disturbing were the figures from Forest County, which showed a decline from 80% support in 1994 to less than 50% support in 1996. "If we can't sell the project to the people in Forest County, where most of the economic benefit will occur, then political support at the state level will evaporate," Alberts reported to management. The moratorium legislation could pass the legislature early in 1997, "making development of the Crandon project impossible." The company must develop alternate or substitute legislation that could sidetrack or delay the moratorium vote. An aggressive program to win back public and political support needed to start immediately.

Rodney Harrill agreed with Alberts. The company's campaign against the moratorium was necessary because of the stakes. "It would probably be the end of Crandon mining and probably the end of sulfide mining in the state," he said. CMC spent about a million dollars in advertisements and lobbying to deflect the moratorium.[6]

Mining supporters outspent opponents nearly 3-to-1 in 1996 while lobbying Wisconsin legislators. The Crandon Mining Company, Wisconsin Manufacturers & Commerce, and the Flambeau Mining Co. together spent $385,800 and 2,391 hours on lobbying. "It was the most expensive lobbying and PR campaign I've seen since I've been in the legislature," said Spencer Black of the mining company's deep pockets. In contrast, mining opponents - mostly the Menominee and Oneida Indian tribes, the Sierra Club and Wisconsin's Environmental Decade - together dedicated $142,591 and 2,413 hours to lobbying. The following year mining interests and their supporters outspent opponents nearly 4 to 1.[7]

During the moratorium debate, the company had purchased "on air" time for a commercial to be televised statewide. It featured Milwaukee union president, Dennis Bosanac of Local 1114 of the United Steelworkers of America. With the union seal in the background, Bosanac said:

Some legislators in Madison want to stop mining. That's like asking over 10,000 working people to stop breathing. Don't they know that thousands and thousands of us work in jobs that depend on mining? Don't they know how important mining has been and will be to Wisconsin? We want to be part of it. Those high-paying jobs belong in Wisconsin, not someplace else.

The televised message reflected the primary political strategy of CMC - obtain the support of organized labor in order to influence the votes of legislators, particularly Democrats.[8]

Wisconsin's labor representatives had privately told CMC they would actively oppose the mining moratorium and publicly support the mine project in exchange for "commitments" from CMC to use union labor in the construction and operation of the project. Building trade members and steelworkers, in particular, sought a commitment that union contractors would be given "preference" in return for their full endorsement of the company's position.

"The steelworkers have proposed a 'neutrality' agreement in which CMC would agree to remain neutral during any organizing and election efforts and the unions would commit to certain ground rules regarding their activities," Rodney Harrill wrote CMC's management committee.[9]

Organized labor unions could make a significant difference in the "dynamics" of the moratorium debate, Dale Alberts advised Harrill in September 1996. The moratorium had polarized the legislature, primarily along partisan lines. "The Democratic Party generally supports Spencer Black and the moratorium. Support for our project (and mining in general) from labor could well mean turning around some of the Democratic votes in the Senate and Assembly."

Milwaukee companies provided many manufacturing jobs, and approximately 10,000 of them derived from mining related projects. Most manufacturing companies were unionized. "If we want grassroots support in key legislative districts, then I suggest to you that there is simply no one better at grassroots organization than organized labor."

Alberts recognized that there was a reluctance on the part of management, particularly Exxon's senior management, to use organized labor in the endeavor. "However, I am confident that we can work with labor without making any expressed or implied commitments regarding the Crandon Project." In addition, wrote Alberts, "I have demonstrable experience in working effectively with labor to assist with highly politicized issues with no expressed or implied commitments."

Alberts thought that at least nine Democratic representatives in the Assembly would probably vote with CMC if labor was engaged. But because management didn't respond aggressively to labor's desire for "preferences" and "neutrality," the company's strategy floundered.[10]

It also floundered because of Gerry Gunderson. He did more to quash the company's labor strategy than anyone else. He became involved in January 1997 after viewing CMC's televised commercial showing Dennis Bosanac endorsing the mine. "I felt that he was not representative of all steelworkers," Gunderson recalled.

Gunderson worked for Rexnord in West Milwaukee and was a member of Local 1527 of the United Steelworkers. After seeing the television ad, he contacted Zolton Grossman and other opponents of the mine who suggested an anti-mining resolution be presented to Gunderson's local union. Gunderson succeeded, and after forming the Committee of Labor Against Sulfide Pollution, he expanded his efforts statewide, winning anti-mining resolutions from local unions and central labor councils in Green Bay, Madison, La Crosse, and Neenah-Menasha. Labor organizations in the Fox River Valley endorsed the moratorium, said Assemblyman Dean Kaufert, Republican from Neenah, because of their members' "avid interest in hunting, fishing and related environmental issues." "A lot of union members were hunters and fishermen," Gunderson agreed. He also told union groups that Exxon and Rio Algom had poor labor and environmental track records in other states and nations.[11]

Then in mid-February 1997, presumably to inform and gain favorable publicity for safe mining, the mining industry sponsored a major conference at the Grand Milwaukee Hotel near Mitchell International Airport. In contrast, Milwaukee Mayor John Norquist criticized the conference and the Crandon mining proposal in a letter. The venture was likely to end up as a "net negative for Wisconsin," and he doubted it could be done without damaging "watersheds, wildlife, forests and tourism....Flushing and diverting mine wastes to our rivers might seem like a good idea to someone on Wall Street, but not to people who enjoy canoeing and fishing on the Wisconsin and Wolf Rivers." Norquist's missive generated more publicity than the participants at the conference.[12]

"Even if the opponents can force passage of the bill, I still believe we will have the votes to further amend it," Alberts hopefully wrote Harrill on March 19, 1997. "In that circumstance, the opposition fervor may diminish and we could focus more attention on the permitting side." After all, the moratorium did not say that the permitting process must stop. "It simply says that DNR shall not issue a permit until an applicant provides the examples of successful sulfide mines."

But hope was fading. The minutes of the company's management committee meeting in Milwaukee on April 24, 1997, reflected the cheerless situation:

* Alliance with Labor not possible.

* Pressure from environmental groups making an impact; affecting moderates, elected officials and government agencies. Moderate people and mainstream groups (such as Ducks Unlimited, Trout Unlimited, Wisconsin Conservation Congress) have sided with anti-mining position.

* Our messages about safety, health and environmental issues are not positively influencing key audiences. Primary impediment is lack of trust....

* Due to low unemployment rates in Wisconsin, the jobs issue is not playing well in advertising.

* General public does not recognize any socio-economic benefits from the project. Jobs issue is not important to the general public.

* Environmental groups continue to be more effective in reaching our key audiences.

* Without labor support it now appears that some type of moratorium bill will be enacted into law.

At times, Alberts seemed almost desperate. Perhaps it would help the company's political situation if they took a fresh look at the design of the tailings management area (TMA), Alberts advised Harrill on March 19, 1997. "At a minimum, I think there are some 'belt and suspenders' type systems that we should consider in an effort to generate a high degree of confidence in the TMA design."[13]

Exxon's ghost haunted the company. In a letter to a Rio Algom executive, Alberts cited a portion of a recent speech given in the state senate by a respected Democratic senator:

"Why should we trust Exxon? What has Exxon ever done to earn our trust? Was it the Exxon Valdez? Was it the artificial manipulation of gasoline prices in the early '70s that resulted in billions of dollars in profits...? I am simply not willing to trust Exxon and we must pass this bill!"[14]

Prospects for the mining moratorium worried Wisconsin's Governor Thompson. "The Governor is with us but getting uncomfortable," said a company report. Harrill met with the Governor on July 22, 1997. In a memo to his file, Harrill described the meeting:

The Governor was sympathetic, but took the opportunity to chastise me for Exxon mismanaging the project by pulling out back in the 1980's against his strong protest. We also expressed our concerns regarding the EPA's role in the process and the Governor referred to Region 5 as "dangerous" and "out of control." The Governor was concerned about the Wisconsin River discharge and the political heat that this has caused....

He asked how we were doing in recruiting labor support and we indicated that a number of labor organizations had entered the debate on our side. The governor said that we need to keep anti-mining legislation off his desk and that he will help us do that.[15]

In mid-October, 1997, more than 1,000 people attended a legislative hearing in Milwaukee to argue the merits of the mining moratorium bill pending before the state Assembly. Charter buses, trucks, and vans unloaded people from across the state, including at least one hundred members of the Menominee Indian tribe and the same number from Walleyes for Tomorrow.

Milwaukee mining equipment companies gave their workers a day off and bused them to the hearing where each worker received a blue T-shirt, a button, and a sign. The T-shirts read "Mining Matters in Wisconsin." "Mining is our state heritage. We're the Badger State," said Tim Sullivan, vice president of marketing for Bucyrus

International, a mining equipment company in South Milwaukee. "Each and every one of us uses an average of 40,000 pounds of minerals a year in our cars, computers, phones and appliances," Sullivan said. "We should remove the badger and the other mining symbols from the state flag if this bill is approved."

Supporters of the moratorium waved homemade signs and sang songs. The "spirit was incredible," said Gerry Gunderson. "It was moving." Yellow barricades separated the two sides, but they managed to jeer at each other. Nine hundred citizens registered to speak, and by the time the hearing ended, nine hours after it began, more than one hundred had spoken.[16]

Early in 1998 the legislature considered the moratorium legislation. Dale Alberts addressed a group of Republican staffers in a legislative hearing room in Madison. He looked and sounded "nothing less than a preacher beneath a revival tent bringing his sermon home. After a dramatic pause, he levels a stern gaze at his audience, and he speaks in a soft, singsong voice," said a reporter. He spoke in front of a table covered with minerals and documents. He and other mining experts spent two hours talking about the Crandon mine, "presenting a dazzling chart and number-filled lecture that Alberts likes to call 'Mining 101.'" He passed around chunks of mined minerals, and talked about tailings and reclamation. Every car had 40 pounds of zinc in it.[17]

But Alberts' efforts couldn't curtail the moratorium's momentum. Environmental groups presented weighty evidence of statewide endorsement - a 2-foot-high stack of petitions with 40,000 signatures. "Here you are," said Keith Reopelle of the Environmental Decade as he handed the petitions to Spencer Black. The 40,000 signatures in favor of the new law were "more signatures than legislators had ever seen for any single piece of legislation," reported the Wisconsin State Journal. "This issue has energized people more than any other issue I've seen in my time in the Legislature," Spencer Black said on January 18, 1998. "We do not want Exxon experimenting with our North woods."

"There is huge support in my district to support the bill. Why would I vote against my people?" asked Shawano Assemblyman John Ainsworth, who had earlier been the target of a recall effort by POW'R for supporting the mine. He had received 2,300 letters,

148

phone calls and other messages from his district - all but a few in support of the bill.[18]

Fox Valley legislators described the moratorium bill as one of the most emotional issues they had witnessed in the Legislature. "Lopsided support for the measure by residents of their districts make their votes all but automatic," said one report.

The old labels of left and right didn't fit anymore. "Caring for the place you're living in and for your kids' future is both a conservative and a radical issue," said one moratorium supporter.

Republican leaders couldn't contain the issue. "We're hoping members will vote with their own colleagues rather than Greenpeace, Earth First! and Spencer Black," said Republican Marc Duff of New Berlin. "I mean who's in the majority here?"[19] But Republican supporters of the mine reluctantly credited Black and environmentalists with a virtuosic public relations campaign. "I think there's a number of members running scared," said Duff. "When you get your buddy down the street who's a fisherman and hunter that's concerned about the issue, it makes you think again."

On January 22, 1998, the Assembly passed the moratorium bill, 75-21, but only after Republicans added a crippling amendment stating only a mine subject to legal or administrative action would be considered "polluting." Therefore, if mines caused extensive environmental damage but were not legally charged, they would not be considered polluting. As Senator Kevin Shibilski of Stevens Point pointed out, the Assembly Bill had "a loophole you could drive a mining truck through."

In early February 1998, Wisconsin's Senate accepted the Assembly bill, but amended it to make it tougher on mining companies. By a vote of 27-5, it required a mining company to cite a mine that had been operated for 10 years and shut down for ten years without causing "significant environmental pollution." This eliminated the need for an official ruling on whether pollution had occurred. A few days later by a 91-6 vote, the Assembly concurred in the change.[20]

Political experts looked at the anti-mining triumph as a model for future citizen-based political actions. "How is it that a diverse group of cash-poor organizers managed to win their battle against a deep-pocketed mining company to pass the mining moratorium law?" The answer was clear. They created a mass movement, energized it, and maintained unity and focus.

Standing on a bank of the Wolf River on Earth Day, 1998, Governor Tommy Thompson signed into law the mining moratorium legislation. Flanked by schoolchildren, business men and women, environmentalists and politicians, Thompson stated, "Nobody in Wisconsin wants a mine that would put our pristine natural resources at risk. If the mine isn't safe, it will not be built. It's that simple and clear-cut." In honor of Evelyn and Roscoe Churchill, opponents of the mine named the moratorium legislation, the "Churchill Mining Moratorium."[21]

*   *   *

Passage of the Mining Moratorium legislation proved to be bittersweet. Although the intent of the bill seemed clear to opponents, the DNR had to interpret and enforce its provisions. Most people assumed the permitting process would be put on hold until the CMC met requirements of the moratorium. After all, Bill Tans had said earlier that if the legislation would be approved, "it is likely the Department would not continue its review of the proposed Crandon project."

DNR questioned how to implement the new law. An internal DNR memo in late October 1998 outlined concerns:

How do we review the material? Contact state/provincial officials, review material for completeness, accuracy, compliance with moratorium criteria, visit the site?

What form does the documentation of our review take? Internal report on each site?....When in the process do we provide a compliance determination for each site?

How do we respond to public requests for the company's material? To requests for our compliance determinations before the Master Hearing?

150

Preparing for the Master Hearing - what if we believe one, several or all sites fail to meet the criteria?[22]

After passage of the moratorium legislation, George Meyer gave a status report on the project to all DNR employees. He realized they had been put in the "difficult position" of defending the Department's handling of the highly controversial proposal, and the staff was "confused" about the Department's position on the moratorium. The Department did not oppose the legislation, Meyer contended, "although we did identify problems in earlier drafts of the bill, many of which were subsequently addressed through amendments." After the Governor signed the bill, the Department would begin drafting "administrative rules to implement the provisions of the bill," said Meyer.

In a letter to Meyer, forty members of environmental groups and twenty-five state legislators urged the DNR to immediately halt the permit process. The process should not move forward until the company complied with the standards of the new law. "As I understand it, a moratorium is a moratorium," said a mining opponent. "That means you stop all work on permitting."

But Meyer soon reversed his stance in a way that seemed biased in favor of the mining industry. In earlier testimony, Meyer had said that a mine would have to meet all standards, but that testimony, Meyer said, had been too quickly prepared by non-lawyers on the DNR staff.[23]

"We made a mistake," he said of his testimony. After further thought, Meyer contended that the law was very clear - separate mines could meet just one qualification. The law required a mining permit applicant to identify a mine that met the 10-year operations requirement. "In a separate paragraph of the law, the mining permit applicant is required to identify 'a mining operation' that meets the 10-year closure requirement without causing pollution." The law was silent on whether a single mine must meet all criteria. "Our interpretation, based on the wording in the law, is that the above criteria could be satisfied by a single example or separate mines."

"If Wisconsin citizens were not convinced before of the DNR's pro-mining bias, this should end all doubt," said David Blouin.

Critics wondered why administrative rules had not been adopted for the law. "You only need to do administrative rules when a law is unclear," said Meyer. Because the mining moratorium law was heavily debated in the legislature, "its key portions are very straightforward," said Meyer. Adding administrative rules could only change the intent of the legislature.

Meyer also contended that the law had no language authorizing DNR to stop its review. The moratorium law stated that the Department could not issue a permit unless the criteria was satisfied. "The moratorium law does not say that the information must be submitted before the Department can review other aspects of the project," the DNR officially declared. "Therefore, by law, the Department will continue processing the mining permit application with or without receipt of the mining moratorium information." Leading up to the Master Hearing, the DNR intended to study the examples submitted by the company and verify whether they did or did not meet the requirements.[24]

*　　*　　*

Critics of the mining project continued to refer to it as the "Exxon mine." A reporter noted that "Anytime Exxon's name began a sentence, someone was sure to finish it with '...Valdez oil spill in Alaska.'" Rio Algom was merely in the background. At the mine, personnel from both companies were cordial and worked well together on many phases of the permitting process. The public didn't know of any divisiveness, and the media didn't report any.

Behind the scenes, however, there was serious tension. Ken Collison, a major Rio Algom executive at Crandon, was not well liked by his Exxon colleagues. "There was a lot of friction between [Collison] and the rest of the team," said an Exxon executive.[25]

With half of the personnel from Texas and the other half from Canada, marrying the two corporate cultures and business philosophies was difficult. Rio Algom perceived Exxon simply as an oil company that knew little about mining. "Rio Algom knew how to build an underground metals mine and Exxon didn't," said one Exxon consultant. For example, in Rio Algom's view the solution for excess wastewater was not to pipe wastewater to the Wisconsin River, but to use grouting to reduce the flow of the water into the mine. Also,

Exxon's senior management in Houston would not acknowledge the risks and lower profits of the mining venture. They "had a hard time understanding the differences between the mining industry and the petroleum industry," said public relations consultant James Wood. The "risk was different."

Exxon was more bureaucratic than Rio Algom and had more layers of management review, but Exxon better understood the regulatory process in the United States than did Rio Algom. The U.S. regulatory process was "very confusing" for the people from Canada, said an Exxon consultant. Exxon was quick to sue Indians tribes; Rio Algom wanted to work cooperatively with them.[26]

Growing frustration with Exxon led Rio Algom to consider buying out their partner. "The key to our strategy is to get Exxon to stand aside now," wrote a Rio Algom executive on March, 26, 1997. "With Exxon in we are virtually certain to lose because their strategy boils down to doing more of the same, which can hardly be expected to produce substantially different results," another Rio Algom official wrote a colleague. Opposition was fast and agile compared to Exxon which was "slow, cumbersome, and bureaucratic." If Exxon could not be persuaded to radically change course, then "I think the outcome is...predictable."

Rodney Harrill of Exxon resided in Crandon, but his wife and family lived in Houston. CMC paid to have Harrill commute every two weeks to see his family. "With his salary and travel expenses, I believe it is costing Crandon over $350,000 per year," Don Cumming of Rio Algom wrote to his file in early January 1998. Since Harrill knew about the negotiations for Rio Algom to buy out Exxon, Cumming wrote, "Rodney's heart cannot be tied to the Crandon Project at present to the same degree it was in the past. I believe he is anxious to get back to Houston and rejoin his family. We cannot afford any lack of pressure on the permitting process. This appears evident in his reaction to the DEIS delay which should have led him directly to the Governor's office." Harrill would need to be replaced.[27]

Exxon's position was they would not accept a buy-out deal unless Rio Algom put the dollars "up front now." Exxon wanted $36 million; Rio Algom offered $25 million. On January 23, 1998, the two companies finalized the sale. In the end, Exxon sold its interest to Rio Algom for $22 million and 2.5% of the profits if the mine ever produced a profit. D.L. McLallen, public affairs manager for Exxon, said the sale was "basically a business decision;" it was only a coincidence that the sale came one day after the Wisconsin Assembly adopted the moratorium legislation. "Believe me," he said, "it was purely coincidental - I had hoped to have different timing."[28]

Rio Algom immediately changed the company's name from Crandon Mining Company to Nicolet Minerals Company (NMC). Don Cumming, an executive vice president with Rio Algom, became the new president of Nicolet, and Dale Alberts its top lobbyist and spokesman.

Nicolet Minerals took its name from Jean Nicolet, the French Canadian explorer who in 1634 became the first non-native to discover Lake Michigan and possibly the first white man to stand in what is now Wisconsin. An interpreter and trader, he spent ten years living among Great Lakes tribes. In its public relations spin, NMC said its name symbolized respect for the environment, respect for Native American people and values, and the close, centuries-old relationship between Canada and Wisconsin.

In 1997 Rio Algom employed 2,451 people, had sales of $1.27 billion, and profits of $57.3 million. Until 1992 Rio Algom had been an affiliate of the world's largest mining company, the London-based Rio Tinto Zinc (RTZ). A report stated:

In 1955, RTZ acquired a substantial interest in several Ontario uranium mines near the north shore of Lake Huron, collectively known as Elliot Lake. During the U.S. nuclear weapons buildup of the 1950s-60s, there were 12 mines in the region employing over 10,000 workers. These mines were combined under Rio Algom mines in 1960, and over the next 30 years were identified with one of the world's most notorious controversies over radioactive contamination of the environment.[29]

In 1985, in an infamous scheme, Rio Algom sold Quebec's Poirier mine for $1 to rid itself of an environmental liability. "These are Canadians dumping on their own neighbors," Al Gedicks charged. "Do you think they will give any more respect to us if they can't respect their own neighbors?" Over the years, the Poirier mine had several owners. "When the last one went bankrupt and couldn't fulfill its responsibility to clean up the site, the Quebec government came to Rio Algom," said a Canadian environmental official. Rio Algom spent millions on the cleanup.

Crandon mine opponents were dismayed that Bill Tans, the DNR regulator responsible for Rio Algom's proposed mine in Wisconsin, knew nothing about the company's polluted Poirier mine in Quebec. "I haven't investigated what they have done outside of Wisconsin," said Tans. "If it happens in Chile or Australia or beyond ten years back, it is irrelevant to us."

Rio Algom made no excuses for its past environmental record. "It does reflect an industry's history, and it is not always very pretty," said Maxine Wiber, Rio Algom's vice president of environmental affairs in Toronto. "The industry recognizes it has an obligation to correct its behavior and correct its past at this site."

One study expressed a broader opinion about Rio Algom's Canadian cleanup saying it fit a common pattern with mining companies throughout the world. "They repeatedly test the limits of environmental laws, and at times go beyond the law, yet want to claim credit when the law forces them to change their practices."

Although environmentalists in Wisconsin castigated Rio Algom's environmental record in Canada, officials in Canada who had regulated the company for many years were more tolerant. "They have got a good corporate track record in environmental affairs," said Bernie Zgola, head of decommissioning for Canada's Atomic Energy Control Board, the agency that regulated uranium mining. Jerry LaHaye, a former district manager for the Ontario Ministry of Environment and Energy, monitored the Elliott Lake mines for the government. "Certainly, there were environmental problems, the result of ignorance in the early days," he said, adding "They have done what I think was a reasonable job to try and protect the environment....I don't know if I would have done anything much differently."[30]

Nicolet Minerals claimed they would work with the public, hold community meetings, and address public concerns. The company would no longer try to convince people that a specific engineering technology or method was safe and would protect the environment. Dale Alberts assured the public the company would listen and work with people "to incorporate changes that the public felt improved the protection of the environment."

The company made three significant engineering changes. It discovered that it could reduce the water inflow to the mine by at least 50% with widespread grouting. This minimized the drawdown to local lakes, streams, and the groundwater table. The company guaranteed the Town of Lincoln that the amount of water pumped from the mine would be less than 600 g/m.

Nicolet Minerals also designed a state-of-the-art water treatment facility that eliminated the unpopular mine water discharge pipeline to the Wisconsin River. When the water went through a double-pass reverse osmosis system, it would be discharged to a soil absorption system and infiltrate back to the groundwater. (The City of Crandon's municipal water treatment plant had used a soil absorption system for years.) The proposal would improve the quality of the treated discharge water and satisfy the public request to "put the water back where you found it," said Alberts.

Finally, to meet objections related to the tailings facility, Nicolet Minerals would add an additional flotation step to the milling process to remove the pyrite. The pyrite would then be mixed with cement and used as underground backfill. In the re-flooded underground, oxygen would be limited so acid production would be minimal. "This ensures the long-term stability of the tailings and alleviates the concern that the pyrite will eventually be released to the environment."[31]

Rio Algom "was more willing to go the extra mile, to spend extra money to address environmental concerns," Alberts proclaimed. "I think we would have designed the best mine in the world." Don Cumming agreed. It was exciting to improve the environmental features of the Crandon mine, he recalled. "We would have the best environmentally designed mine in the world....That would be the target that everyone would have to aim at from then on."

Despite the proposed changes, the Milwaukee Journal-Sentinel questioned whether such changes were sufficient to protect the environment. "This is still a lousy place to dig a mine. The jobs the mine would generate are just not worth the environmental risks or the potential harm to the area's tourism and recreation industry."[32]

Nicolet Minerals tried hard to appease the three Indian tribes, particularly the Mole Lake community. The company's efforts focused on gaining trust and respect from tribal members one at a time. The company contributed to a Native American scholarship fund at the University of Wisconsin and helped sponsor a National Spirit Award dinner that benefited Native American education.

When the Sokaogon Chippewa set their strict non-degradation standard on all water resources at Mole Lake, the company could have sued and sought a court ruling on the validity of the standards, or they could design the project to comply with Mole Lake's standards. The company chose the latter option. "We knew one thing - fighting in court over this issue would cost time and money," said Alberts; "it could also cost us much more - an amicable relationship with our Native American neighbors." The only way the company felt comfortable with Mole Lake's action was to comply with the non-degradation standards of the reservation "so we have no measurable impact on tribal water quality."[33]

But no deep rapport was achieved. The company brought in experts on growing rice and tried to explain the ideas to Mole Lake. "They were firmly convinced...that we were liars," said Don Cumming. "It didn't matter what we said."

"Requests for meetings were rejected," said an internal company memo. "Chairman McGeshick stated that even if he wanted to meet with NMC, the Tribal Elders would not allow it. NMC continues to communicate with the community through direct mailings and individual meetings."[34]

By 1998, reinforced with gaming revenues and determination, the Potawatomi became the dominant tribe leading the opposition. They hired high-end attorneys and consultants. While Nicolet scaled back its budget, the Potawatomi were building a powerful record against the mine. "A lot of what they submitted was so credible and solid, it went unrebutted," said an Exxon consultant. Lawyers with large companies like Exxon often overpowered opponents with a mountain of paper. "It happened to Nicolet Minerals in reverse," the consultant observed. "The [Potawatomi] did it to Nicolet with just a mountain of paper. What an irony."[35]

CHAPTER EIGHT

NEW OWNERS, SAME FRUSTRATION

Up to this point in time, the Crandon Mine controversy had spanned a generation. "In Nashville, a town of 1,157," reported the Chicago Tribune, "the dispute has so severely strained relationships that the township is threatening to break in two. Generally, the south part of Nashville - where many former city dwellers have built homes around lakes - opposes the mine. The north end of the township - which has deep roots and is largely rural and working-class - supports the mine."

Mine opponents pressed on and launched a new issue that would complicate mining. Spencer Black introduced legislation to ban the use of cyanide in the mining process. He cited a recent cyanide spill in Romania that contaminated the Tisza and Danube Rivers, as proof that the ban was needed. Proponents of the cyanide ban were troubled not only about lingering pollution from cyanide near the mine site but also about the danger of shipping the cyanide.[1]

Projections estimated between seven and eighteen tons of sodium cyanide per month would need to be trucked to the Crandon mine site. Critics claimed that communities along the truck routes were horrified at the potential for a spill. "We all know too well that truck is going off the road," said Potawatomi Tribal Chairman Gus Frank.

Dale Alberts contended cyanide was used in other industries throughout the state, by fifty different companies in seventeen other counties. It was unfair to target only the mining industry. Critics responded that cyanide use by other industries in Wisconsin was not on the same scale as the amount proposed for the Crandon mine. They claimed that records from the National Toxic Release Inventory indicated that the largest use by any other Wisconsin industry was only twenty-four tons a year.[2]

Most of the cyanide would be consumed in the process that separated zinc from copper, refuted Alberts. "Cyanide would not be detectable in the wastewater or leftover ore generated by the mining operation. There's no technical or environmental merit to the bill. The proponents have seized on the use of one chemical in an attempt to frighten people." The proposed legislation was not about the environment. "It's not about sodium cyanide. It's about politics. And it's another spurious attack on the Crandon mine."

The DNR agreed that cyanide was not a danger. From the DNR's standpoint, Bill Tans said, "Cyanide is not considered one of the major environmental problems." The cyanide use planned for the Crandon mine site was different than that used in Romania or western gold mines. The proposed ban on cyanide didn't distinguish between two different mining methods. With open-pit mining, cyanide was difficult to control, but the flotation method that Nicolet planned to use, where the chemical was diluted, presented little risk of dangerous cyanide being released into the environment.[3]

Nevertheless the Wisconsin Conservation Congress, whose mission is to advise the Department of Natural Resources on policy issues, twice voted overwhelmingly to support the ban on the use of cyanide in mining, and many local governments, conservation groups and unions passed similar resolutions.

When a State Senate committee recommended passage of the legislation to ban the use of cyanide in mining metals, Dale Alberts exploded. He angrily responded that it was "just dirty Spencer Black politics....It is nothing but a cheap Spencer Black attempt to stop our mining project. They were not banning the use of cyanide for any other industrial facility in the state," Alberts charged. "Just mining." There was no adequate substitute for sodium cyanide. "I think that would be a problem for the company," Tans said.

In early November 2001, the Wisconsin Senate, by a vote of 19-14, passed legislation to prohibit the use of cyanide and cyanide compounds in mining. The Assembly and the governor still had to approve the measure before it became law.[4] Meanwhile, the DNR and the mining company still disagreed on computer models created to register the impact the mine would have on groundwater. "They have their model and we have our model," Bill Tans said. "We've sort of gone our separate ways. Evaluating that will take a number of months," Tans added. "The upshot of all this is that we don't know for sure when we will have a draft environmental impact statement.

We've always said it will happen by such and such a day. But then it never happens."

Finally the DNR and Nicolet agreed that the company would limit the flow of water into the mine at 600 gallons per minute. That amount would protect the surrounding wetlands, lakes and rivers from damage. "It's an important step," Tans said.

"The state Department of Natural Resources might be within months of issuing its long-awaited draft environmental impact statement," reported the <u>Wisconsin State Journal</u> on October 1, 2001. "Computer models showing the impacts of the mine on groundwater, as well as on nearby lakes and wetlands, are nearly complete," said Tans. He predicted that the draft environmental report could be out by the spring, 2002. "Things are going well," Alberts said. "We're encouraged."[5] But the euphoria wouldn't last.

Changes in corporate ownership had occurred. In August 2000, Billiton, a South African mining company, purchased Rio Algom and inherited the Crandon Mine. Billiton thought its acquisition of Rio Algom would provide a significant entry point into the global copper business and be a vehicle for future development of other base metals opportunities and would bring to Billiton a quality portfolio of "low-cost, high-growth operating assets and development properties." Billiton employed 32,000 people worldwide and had a market capitalization of about US $10 billion. A leading mining and metals business, the company had operations in Australia, Brazil, Canada, Colombia, Mozambique, South Africa, and Suriname.

Opponents of the mine quickly organized a campaign to contact Billiton officers and directors. "A listing of the Billiton directors is attached and we encourage you to contact these individuals representing your own organizations. Provide these officers and directors with letters and other documents demonstrating the breadth of the opposition to this mine in Wisconsin and encouraging them to 'write off' this economic drain on their portfolios."[6]

While writing a contact in South Africa, Zolton Grossman made an attempt to enlist an international hero in the effort to block a mine near tiny Crandon, Wisconsin. "Are there other groups around the world that have opposed Billiton, and how do we reach them? Do you have any suggestions on how to get [Nelson] Mandela to write Billiton?"[7]

Billiton's Senior Corporate Affairs Manager Marc Gonsalves responded to critics, saying his company did not want to be where they were not wanted. The company was considering the Crandon mine "very carefully" because it had received an "endless stream of e-mails" from Wisconsin and around the world. Billiton owned the Crandon Mine for a brief period only.

On May 19, 2001, the Australian mining giant, Broken Hill Proprietary (BHP), hosted its Annual Meeting at its headquarters in Melbourne, Australia. The major agenda item was consideration of a merger with Billiton. As shareholders entered the meeting at the Melbourne Concert Hall, they encountered 2,000 protesters, among them members of the environmental group, Friends of the Earth Melbourne, who brandished signs reading "Stop Billiton's Crandon Mine in Wisconsin, USA," "Protect Indigenous Rights in Wisconsin, USA: Drop the Crandon Mine," and "Billiton's Nicolet/Crandon project: A Disaster for BHP in USA."

As the name changed, the controversial issues remained. What started as the Crandon Mining Company in the 1980s (Exxon Coal and Minerals Company) changed to the Crandon Mining Company (Exxon Coal and Minerals Company and Rio Algom). Next it became the Nicolet Minerals Company (Rio Algom), then Nicolet Minerals Company (as a subsidiary of Billiton), and now it was the Nicolet Minerals Company (as a subsidiary of BHP Billiton). Despite ownership changes, opponents simply referred to the mine as the "Crandon Mine."[8]

In an 11,000-mile journey to Johannesburg, South Africa, a Mole Lake delegation journeyed to the World Summit on Sustainable Development to meet with executives at BHP Billiton. "It's the little guys, from one of the smallest and poorest tribes in the nation, going all the way to Africa,...to talk with a giant mining company," said the tribe's attorney, Glenn Reynolds. "I think it's going to be a great moment. Mainly, we want to let the corporation know where we're coming from. And we want to encourage them to sell."

On August 30, 2002, the Sokaogon delegation with their environmental experts met with Brian Gilbertson, BHP Billiton's

CEO, to discuss the Crandon proposal. The meeting that was scheduled to consume one hour ran for almost two and one-half hours. The Sokaogon spiritual leader, Robert Van Zile, began the meeting with a prayer and a pipe ceremony. "The 'no-smoking' ban in the corporate boardroom was temporarily lifted to allow the ceremony to take place in which the BHP Billiton executives fully participated," Reynolds recalled.

The results of the meeting encouraged the delegation who thought company executives were malleable. A representative of the tribe said, "I think the company executives could see that mining in this area is completely incompatible with the concept of sustainability and respect for the integrity of indigenous cultures."[9]

Financial problems continued to mount for Nicolet Minerals Company and BHP Billiton. In 2001-2002 zinc and copper prices reached historic lows. From December 2001 - February 2002, BHP Billiton reported a 33% drop in profits. "Even if a Crandon permit were issued, the depressed state of metals prices has led to the closure of other mines around the world," stated one report. "The proposed Crandon mine is not economically viable."

In early September of 2002, the DNR rejected one of three mines that Nicolet had submitted as examples of non-polluting mines. The Sacaton Mine, an open-pit copper mine in Arizona, did not comply with the moratorium law. In addition, the Mining Impact Coalition of Wisconsin researched three mostly copper mines owned by BHP Billiton in Arizona and Nevada, and claimed there were 31 spills of hazardous materials over four years. "Nicolet Minerals has never operated a mine. But its parents have, and one of its parents, BHP, in a four-year span at three mines in the United States seems to be a chronic polluter," Dave Blouin charged. Dale Alberts weakly responded that chemical spills were not unusual at any large-scale industrial activity.

Nicolet Minerals Company "was much easier to deal with," said the DNR's Stan Druckenmiller. "They quickly responded to everything we put forward to them." Unfortunately, Nicolet didn't return the compliment.[10] Except for the time when both sides expressed optimism about the permitting process, the DNR frustrated Nicolet. On one occasion, the DNR told NMC that it needed access to the mining site to reassess "what is a lake." The company granted the access, but strongly objected to the incredible request. The Crandon Project site was probably the most exhaustively investigated area in Wisconsin's history. "Your agency first began studying the site over 20 years ago," replied Gordon Reid, "and has spent many millions of dollars and many thousands of hours over the past generation investigating the lakes, ponds, streams, wetlands, and dry land in the vicinity of the proposed mine. NMC cannot believe that the Department could somehow have 'overlooked' the existence of an additional 'lake' over these many years."[11]

The company was spending millions of dollars on the project and thought it deserved quicker responses from the DNR. "At what stage does the analysis of minutia stop?" Don Cumming demanded of George Meyer. There seemed to be no end to the information requested by the DNR. "I cannot understand why the DNR wants to micro-analyze this project."

"It had been studied to death," Dale Alberts recalled, "but the DNR treated this like it was a totally new project. Much to our chagrin, the DNR would throw another letter our way with 30, 40, or 50 more questions that had to be answered."[12]

"We believe that the plan provides the basis to limit water inflow into the proposed mine," the DNR's Carlson wrote the company. "However, we have identified several issues that we believe need additional clarification...."

As late as July 2001, according to the DNR, several critically important work items remained to be finished before the department completed its permit reviews. The list included:

* Complete groundwater flow modeling and mine inflow estimates at the proposed project site;

* Provide final comments on the company's surface water mitigation plan;

* Conduct final review of tailings management area waste chemistry;

* Complete analysis of tailings management area impacts to groundwater quality;

* Review re-flooded mine water chemistry;

* Review re-flooded mine impacts to groundwater quality;

* Review the company's plan for its irrevocable trust; and

*Complete preliminary review of mining moratorium candidate mine sites.[13]

"How much more can you do?" asked mine supporters. "If you can't get it done in seven years, how long is it going to take you?" Crandon Mayor Pat DeWitt asked. The mining permits were taking longer than NASA in the 1960s needed to land a man on the moon.

Future mining projects across Wisconsin were at stake. If the Crandon mine was successful, it would show that companies actually could secure a mine permit in the state. "If they could get it permitted, other companies would be interested in exploring," said Laura Skaer, executive director of the Northwest Mining Association. "All those people who have homes on the lake, computers, cell phones, SUVs, they all have stuff that came out of holes in the ground," Skaer added. "Unfortunately the minerals are where God put them. It's not like building a Ford plant."[14]

In mid-December 2002, Dale Alberts bitterly complained to the DNR. The department's staff had repeatedly informed NMC that they were nearing the end of their review and that a Draft Environmental Impact Statement would be forthcoming in the near future.

NMC has been hearing this same communication since 1999. Yet, it is now the end of 2002, four years after our documents were submitted, and the DNR has yet to "officially" declare that their review is complete. Keep in mind that this is for a project that was nearly permitted in the 1980s and has been improved dramatically with the revisions incorporated into the proposed project in the 1990s. NMC is thoroughly disappointed and outraged with the DNR's lack of progress towards issuance of a DEIS. In fact the DNR's lack of performance is forcing us to conclude that the DNR may lack the will or the ability to complete this process....

The DNR and their consultants have demonstrated a history of delay, mismanagement of the review process, an inability to complete their reports in a timely fashion, and contempt for adhering to any type of schedule for bringing this process to conclusion.

The DNR created a "moving target," said Don Cumming. "You solve a problem. And as soon as you solve it, it's not good enough. You just keep working, working and working."[15]

For a period of time in the summer of 2002, the state of Wisconsin considered purchasing the mine site. BHP Billiton would have benefited substantially by selling the mine to the state, but no deal was reached. Governor Scott McCallum's interest waned when he learned that appraisals of the mine's value came in at $51.2 million to $94 million.

In mid-September, 2002, BHP Billiton laid off eight of its remaining nine employees at the mine site. Paul Benson, a vice president at BHP Billiton, said the company needed to cut costs at the Crandon site and would try to sell the mine. Opponents were delighted. "It's a dead project," said Al Gedicks.[16]

The lone remaining mining official, Nicolet Minerals President Dale Alberts, claimed the project was not dead, only in need of a new owner. The permitting process would continue. "I will be around to manage that process and the sale process," said Alberts. "We'll see who else is interested in this project....It's a good asset and should be developed."

"It's unfortunate," Alberts added. "Clearly it's a disappointment for our team. We've had a full staff since last December with nothing to do. The company cannot continue to pay people to sit around doing nothing waiting for the DNR to do its job."[17]

Only a half-dozen mining companies were capable of investing the $350 million needed to construct the Crandon mine. It was an attractive ore body, but whether any company "wants to bang their head against the Wisconsin brick wall is another story," said Laura Skaer. "It is pretty clear to me that there is a strong segment of the Wisconsin population that doesn't want this mine built."[18]

CHAPTER NINE

SOLD TO THE INDIANS

In April 2003, Northern Wisconsin Resource Group (NWRG), a newly formed company controlled by Gordon P. Connor, president of Nicolet Hardwood Corporation of Laona in Forest County, bought the mine property. Connor wanted to hold the land and keep the deposit intact, hoping it would eventually be developed. Connor stated that the underground mining of zinc and copper could be conducted in an "environmentally safe manner," and his company would generate a $5 million annual payroll and provide two hundred local jobs.

Connor's acquisition did not appease opponents. "Different player, same story," said Joan Delabreau, chairperson of the Menominee tribe. "It is a big threat to the environment, and we are downstream. It is vital to us that it not succeed."[1]

The Connor family had been involved in the logging and lumber business in northern Wisconsin for many years but had never owned a mine. The family had a track record of being good stewards of the land. Hoping to preserve the environment and mollify opponents, the company planned to simplify the mining process by eliminating the tailings ponds and transporting the ore by rail to Northern Michigan or Canada as did the Flambeau mine at Ladysmith, Wisconsin.

Gordon P. Connor's son, Gordon R. Connor, 37, was the mine project manager. He said the family had no illusions that they were miners. "We are not proposing that a lumber company is going to mine it, nor are we proposing that we have the kind of resources to pour into a mining project." The new company claimed it had "serious conversations" with two excellent mining companies.[2]

Shortly after his company purchased Nicolet Minerals, Gordon P. Connor requested the DNR to temporarily stop the permitting process. Later he reversed his request, asking that permitting resume. The senior Connor later complained that he experienced nothing but frustration in dealing with the DNR. "Underlings" in the DNR said that "our simplified" plan would have to be "restudied." He felt that the DNR had an imbedded anti-mining group.[3]

Glenn Reynolds now assumed a critical role in the Crandon controversy. He had graduated from the University of Wisconsin Law School in 1977 and practiced law in Madison. Following his wife's death, Reynolds reviewed his career plans and realized he was tired of civil and criminal law. He wanted to be involved in something "bigger." In 1996, with the Crandon controversy raging, he volunteered to work "pro bono" for the Sierra Club and for the open meetings case in Nashville. Two years later the Mole Lake tribe hired him as their lawyer.

After the failure of BHP Billiton's negotiation with the state of Wisconsin, Reynolds, representing the tribe, wanted to talk to the company about purchasing the property, but BHP Billiton sold it to Connor instead. "When I heard that news I was devastated," recalled Reynolds. He assumed the Connor family would secure backing from another mining company. "I was so dejected and so were all the [Mole Lake] Council members."

But the Connor family made little headway with their mining plans. When they took over the property, "the burn rate" was $400,000 a month. "I quickly reduced it to about $200,000," said Gordon P. Connor, but that amount was still too much. No buyers had come forth to purchase the property, and since the Connors judged the DNR as uncooperative, "We had to reassess the situation."

In the summer 2003, the younger Connor tried to contact the Mole Lake Sokaogon about selling Spirit Hill to the tribe. Spirit Hill occupied 320 acres of the 5,000 acre property. The property was sacred to the tribe because the area contained the graves of more than 500 Chippewa and Sioux warriors who had died in the 1806 battle for control of the wild rice beds. For the company, the sale would reduce their monthly losses, but as usual, the tribe refused to talk.

"We want to be sensitive to their spiritual needs," the younger Connor told the <u>Green Bay News-Chronicle</u> on July 30, 2003. "We'll give them right of first refusal, but we're in the active process of selling Spirit Hill, and when it's done it's no longer our concern."[4]

Sandra Rachel, Mole Lake's tribal chairperson, read the Chronicle article and informed Reynolds. "I called Sandy back and said 'I don't know what this means, but things are pretty volatile right now. I think I should call Connor.'" With the support of Rachel, but unknown to Mole Lake's Tribal Council, Reynolds made an appointment with the younger Connor. "I stuck my neck out on this one," Reynolds recalled. "I was on the authority of the chair [Rachel]; not the council. That can be a dicey proposition, but I didn't think we had time to wait." He took along Roman Ferdinand to witness the discussion. On the way to the meeting, Ferdinand asked "Am I going to get fired for this?"

The pair spent several hours with Connor at the Nicolet Minerals office in Crandon. After listening for a while, Reynolds asked Connor, "Why pour more money down the hole? Why sell just one piece? Why not sell the whole thing?" Reynolds anxiously waited for Connor's reaction. Connor gazed at the ground, then looked up and said, "Well, anything is for sale."

After the meeting Reynolds reported to Mole Lake's Tribal Council. "They were annoyed with me, but they were also excited." After receiving authorization from Mole Lake's Tribal Council to offer $12.3 million, Reynolds arranged a meeting with Gordon P. Connor. "We didn't have the money," Reynolds conceded.

At the meeting Gordon P. Connor was stunned. "I didn't think they had the capacity to do that [on their own]." But "[Mole Lake] assured us that they did. We put two and two together and figured out that they had worked out some arrangement with the [much wealthier] Potawatomi."

Reynolds had indeed been talking with the Potawatomi, but negotiating with their "seven lawyers"' was frustrating. "That negotiation was a helluva lot harder," Reynolds recalled. "The tribe with the money had all of the lawyers." According to Reynolds, the Potawatomi wanted the property, except Spirit Hill. "They almost tripped up our negotiations several times on purpose."

The best way to convince the Potawatomi was to bypass their lawyers, Reynolds reasoned. He drafted a letter for Sandra Rachel to send to Gus Frank, chairman of the Potawatomi, suggesting a council-to-council meeting. "That was the breakthrough," said Reynolds. The councils met and within fifteen minutes agreed to purchase the property and divide the land.

The final purchase price was $16.5 million. The Potawatomi easily raised their portion, $8.25 million, but the impecunious Mole Lake tribe had to figure out a way to come up with their share of the money. In the process of negotiating with the Connor family, Reynolds discovered that BHP Billiton had loaned the Connors $8 million for a 3-year term with no interest. The loan could be assigned to the Mole Lake purchasers, and they would pay $250,000 to meet the terms of the purchase agreement.

Gordon R. Connor subsequently complained that Wisconsin's "anticorporate culture" defeated the mine and felt the anti-mining message was clear. Dale Alberts said that the Crandon mine "is dead and gone forever. I think it is essentially the end of mining in the state. It is a bitter pill."[5]

Wisconsins's DNR was also disappointed. "What was frustrating to the [DNR] staff," said George Meyer, "was there was no conclusion or final assessment." Larry Lynch agreed. "I'd been working on this project for almost twenty-five years. You'd like to see some conclusion."

Mine supporters were critical of the unwelcoming business climate for mining in the state. Conservative Milwaukee talk-show host, Charlie Sykes, contended that the mining companies spent millions of dollars trying to mitigate any effects the mine might have on the environment, but it was never enough.

Over the past decade, [the mining company] has been the recipient of a litany of governmental inefficiencies and has become emblematic of a bureaucratic process that is stifling the economic infrastructure of the state.

It is a sad commentary on the business climate in Wisconsin when every major mining concern in the world that we engaged said repeatedly "not interested in doing business in Wisconsin." Wisconsin's political and regulatory process is out of step with the state's economic needs.

An editorial in the Wisconsin State Journal criticized the DNR for the "outrageous runaround" it gave the mining companies:

To be sure, there are complications to the story. The mine development company had four successive owners over time; technology changed, plans were withdrawn and altered, new questions arose. Those are all facts that regulators and lawmakers had to cope with. But when the complications are used to justify 21 years without a decision, they are excuses. And that is inexcusable.

All this created the impression that Wisconsin abused its regulatory process for the sake of politics. Officials couldn't find a way to reject the mine proposal, so to avoid a political firestorm, they simply made sure the process would go on endlessly. No wonder all the mine site owners gave up in frustration.[6]

After the final sale papers were signed, Mole Lake tribal members crowded the street outside the headquarters of Nicolet Minerals in Crandon. They slapped a huge "Sold" sign across the building. In impromptu ceremony, tribal members pounded a drum. The beats reverberated as people hugged and cried tears of joy and relief. The purchase protected the Wolf River, the wetlands and the groundwater, said Gus Frank. "It ends the threat to the tourism economy - the economy that most of us in northern Wisconsin, including the tribes, depend on."

There were additional celebrations. At one, Tina Van Zile and Roscoe Churchill jointly cut a "WE WON" cake. Mine opponents were photographed in front of Spirit Hill. A formal celebration was held in Green Bay on a Saturday afternoon in the Brown County Veteran's Memorial Arena where a coalition of thirty-three environmental groups honored the Forest County Potawatomi and the Mole Lake Sokaogon Chippewa tribes for buying the land and ending twenty-eight years of acrimony.[7]

Opponents had fought the mine, using diverse ways and means over many years. They voted out the entire board of the Town of Nashville, sued public officials for open meetings law violations, sought to enforce groundwater laws, lobbied for the mining moratorium, and tried to ban cyanide in the mining process.

Much of the effort was conducted at kitchen tables in little houses. "This is democracy at its best," George Rock observed. "You have to understand what we were up against. We were naive enough to think we could win," Rock concluded, that "we could beat money."[8]

After the two tribes purchased the mine, the Sokaogon Chippewa were hard pressed to pay their $8,000,000 debt to BHP Billiton. "These people have mortgaged their homes and their futures and probably their children's and grandchildren's futures," said Bob Schmitz, a leader of the Wolf River Watershed Alliance.

In May 2006, BHP Billiton essentially forgave the debt and stated the Mole Lake tribe would be the beneficiary of a trust fund established by BHP. The Madison Community Foundation was named trustee of the $8 million endowment. As sole beneficiary, the Mole Lake tribe would receive the fund's $400,000 to $600,000 annual earnings. The earnings could be spent on health care, the environment, education, housing, tribal courts, and water and sewer improvements.

Tracey Whitehead, speaking for BHP Billiton, said the action gave her company a "successful and permanent" exit from the project while contributing ongoing benefits to the tribe. "It is a gift. It is basically a gift out of the blue," said Tony Phillippe, administrator of the Mole Lake Sakaogon band, and it left tribal leaders "flabbergasted."[9]

*   *   *

Journalist Ron Seely of the <u>Wisconsin State Journal</u> had covered northern Wisconsin Indian topics for twenty-five years and noted a major transformation. "In 1978, the reservations were places where alcoholism was epidemic, early deaths were common, and government handouts were the rule." The treaty rights victory of the 1980s began a change. Enduring taunts and racial insults, the Indians stood together on the spear fishing issue. A quiet pride emerged. "By making their stand, they were following ancestors who had fought years before to carve out and keep their cultures alive and to retain at least small pieces of their ancient homelands."

More positive changes followed. Many were supported by the influx of dollars earned from casinos. "When visiting the reservations twenty years ago, it was rare to come across indigenous language classes or cultural teachings. Now they are the norm. Attend a tribal powwow sometime and notice the strength and the pride in the faces of the children."

Tribal leaders' pride grew when they announced the purchase of the Crandon mine. In the deal they struck, the tribes showed they were now an economic and political force, and "the strength and the will and the spirit of their ancestors burns strong in them, too."[10]

For two decades DNR scientists conscientiously studied the mine and its possible effects on the environment, and often their motives were understandably questioned. It appeared at times that the DNR abused the regulatory process. They couldn't find a reason to reject the mine, didn't want to create a firestorm by approving it, and therefore let the process go on interminably.

The exceptionally slow pace of the DNR's permitting review and the demanding nature of that review frustrated mine owners. However they, too, made some poor judgments, the most notable being Exxon's decision to pull out of the permitting process in 1986 and the attempt to build a long pipeline into the Wisconsin River to dispose of wastewater. The company could never guarantee, without reasonable doubt, that the mine would not harm the environment.

For over twenty years mining opponents were determined; they persevered. Their extensive and unified grassroots campaign combining petition drives, resolutions, educational forums, lobbying, technological innovations, and tens of thousands of letters and e-mails was democracy at its best.

ENDNOTES

Abbreviations for Notes

GRP   George Rock Papers, White Lake, Wisconsin

MCT   Madison Capital Times

MJ   Milwaukee Journal

MJ-S   Milwaukee Journal-Sentinel

MS   Milwaukee Sentinel

NMCP   Nicolet Minerals Company Papers, Crandon, Wisconsin

OH   Oral History

SWP   Sonny Wreczycki Papers, Pearson, Wisconsin

WDNRP   Wisconsin Department of Natural Resources Papers, DNR Office, Madison, Wisconsin

WDNRP/SHSW Wisconsin Department of Natural Resources Papers, State Historical Society of Wisconsin, Madison, Wisconsin

WSJ   Wisconsin State Journal

Introduction

1.   <u>Final environmental impact statement, Exxon Coal and Minerals Co. Zinc-Copper Mine, Crandon, Wisconsin</u> [1986], 113-114, WDNRP.

2.   <u>MCT</u>, July 16, 1977; Nesper, <u>Mushgigagamongsebe District</u>, 17, 19; <u>Final environmental impact statement, Exxon Coal...Wisconsin</u> [1986], 113, WDNRP.

3.   <u>Final environmental impact statement, Exxon Coal...Wisconsin</u> [1986], 115-116, WDNRP; <u>Proxy Issues Report</u>, April 28, 1983, GRP; OH, Barry Hansen, OH, Daniel Polar, OH, Sylvester Polar.

4.   <u>Final environmental impact statement, Exxon Coal...Wisconsin</u> [1986], 205, WDNRP; <u>MJ</u>, November 24, 1983; <u>Wisconsin Report</u>, December 20, 1983.

5.   <u>WSJ</u>, June 14, 1976, and October 9, 2003; Nesper, <u>Mushgigagamongsebe District</u>, 3, 27-28, 34.

6.   <u>MCT</u>, June 21, 2003; <u>WSJ</u>, October 9, 2003; OH, Fred Ackley.

7.   Nesper, <u>Mushgigagamongsebe District</u>, 25, 29.

8.   <u>MJ-S</u>, May 14, 1995; <u>Wisconsin Report</u>, December 20, 1983.

Chapter 1 Reaching Consensus

1.   "#31. Case History of Metallic Mineral Exploration in Wisconsin, 1955 to 1991," NMCP; <u>Extracts</u>, Exxon Minerals Company Newsletter, September/October 1985, 3; <u>Wisconsin Engineer</u>, April 1997; OH, Edwarde May.

2.   "Zinc Facts," WDNRP; <u>Business Week</u>, March 28, 1977; <u>Green Bay News-Chronicle</u>, May 10, 2000.

3.   Clipping [1979], GRP; MJ-S, February 23, 1997; New York Times, November 26, 1976; WSJ, May 15, 1976 and June 14, 1976; Gough, Farming the Cutover, 18-20, 226; Kates, Planning a Wilderness, xiv-xvii; Thompson, History of Wisconsin, 237-238

4.   MCT, July 9/10, 1994; Stevens Point Journal, February 18, 1981; WSJ, May 15, 1976.

5.   "A Socio-Economic Impact Study of the Proposed Crandon Mine Project on the Menominee Indian Community," Menominee Treaty Rights and Mining Impacts Office, 1997, WDNRP; Clipping [1979], GRP; Appleton Post-Crescent, September 27, 1993; MJ, April 11, 1978; Threinen, "Fabulous Wolf," 3-5.

6.   Stevens Point Journal, February 18, 1981; This is Exxon, 4.

7.   WSJ, June 29, 1976; OH, J. Wiley Bragg, OH, Edwarde May.

8.   MCT, July 11, 1977; MJ, August 11, 1976; New York Times, November 26, 1976.

9.   MCT, June 29, 1977; MJ, May 16, 1976.

10.   Clipping, June 1982; Green Bay News-Chronicle, May 10, 2000; OH, Arlen Christenson.

11.   Clipping, GRP; Peshek, "New Metal Mining in Wisconsin," 24; Peter Peshek, "The Environment and the Engineer," speech, September 1982, Peter Peshek Papers; Huffman, Protectors of the Land, 151.

12.   Bill Tans to Steve AveLallemant, February 26, 1997, WDNRP; MJ, October 24, 1982; Evans, "Development of Metallic Mining Policy in Wisconsin," 100; Peshek, "New Metal Mining in Wisconsin," 24; OH, George Rock.

13.   DLAD newsletter, February 1980, 1, 4; MJ, October 24, 1982; MS, June 4, 1979; OH, Tom Dawson, OH, Peter Peshek.

14.   Peter Peshek, "Environmental Decision Making in the 1980s: Consensus or Conflict," speech, August 18, 1982, Peter Peshek Papers; "#31. Case History of Metallic Mineral Exploration in Wisconsin, 1955 to 1991," NMCP; Steve Brick, "Mining Policy in Wisconsin," April 26, 1982, 15, 23, paper in Mary Lou Munts Papers; Wall Street Journal, January 4, 1978.

15.   Clipping, fall, 1979, Wisconsin Legislative Reference Bureau Papers; Business Week, March 28, 1977; Evans, "Development of Metallic Mining Policy in Wisconsin," 95.

16.   Douglas Amy, "Potential Political Problems in Wisconsin's 'Consensus Process,'" March 27, 1985, 3, paper in the Mary Lou Munts Papers; Steve Brick, "Mining Policy in Wisconsin," April 26, 1982, 15-20, paper in the Mary Lou Munts Papers; "Wisconsin's Net Proceeds Tax on Mining and Distribution of Funds to Municipalities," April, 1997, WDNRP.

17.   Mary Lou Munts to Mary Grady, August 27, 1984 and March 26, 1985, Mary Lou Munts Papers; League of Women Voters Questionnaires #1 and #2, Munts Papers.

18.   James Derouin, speech, "The Wisconsin Model - A Consensus Approach," May 1, 1981, Peter Peshek Papers; Tom Evans interview with Mary Lou Munts (1990), Mary Lou Munts Papers; OH, Linda Bochert, OH, Kathleen Falk.

19.   James Derouin, speech, "The Wisconsin Model - A Consensus Approach," May 1, 1981, Peter Peshek Papers; OH, James Derouin.

20.   Douglas Amy, "Potential Political Problems in Wisconsin's 'Consensus Process,'" March 27, 1985,

paper in the Mary Lou Munts Papers; Evans, "Development of Metallic Mining Policy in Wisconsin," 99-100; Tom Evans interview with Mary Lou Munts (1990), Mary Lou Munts Papers; OH, Tom Dawson, OH, James Derouin.

21.   Evans, "Development of Metallic Mining Policy in Wisconsin," 98-100; Peshek, "New Metal Mining in Wisconsin," 26; Daily Cardinal, February 14, 1980.

22.   "#31 Case History of Metallic Mineral Exploration in Wisconsin, 1955 to 1991," NMCP; Evans, "Development of Metallic Mining Policy in Wisconsin," 62, 103-104; Russell, "Exxon Crandon Zinc-Copper Project," 12.

23.   "#31 Case History of Metallic Mineral Exploration in Wisconsin, 1955 to 1991," NMCP; Evans, "Development of Metallic Mining Policy in Wisconsin," 101-102; Russell, "Exxon Crandon Zinc-Copper Project," 12.

24.   Bill Tans to Sara Johnson, November 7, 1995, WDNRP; Evans, "Development of Metallic Mining Policy in Wisconsin," 105; James Derouin, speech, "The Wisconsin Model - A Consensus Approach," May 1, 1981, Peter Peshek Papers; "Testimony of Gerald D. Ortloff Before the Wisconsin Metallic Mining Council," October 20, 1980, Peter Peshek Papers; OH, Tom Dawson.

25.   Evans, "Development of Metallic Mining Policy in Wisconsin," 108; Peshek, "New Metal Mining in Wisconsin," 25.

26.   Russell, "Exxon Crandon Zinc-Copper Project," 13; OH, James Derouin, OH, James Klauser, OH, Robert Russell.

27.   Evans, "Development of Metallic Mining Policy in Wisconsin," 106; "Wisconsin's Net Proceeds Tax on Mining and Distribution of Funds to Municipalities," April, 1993 WDNRP; Russell, "Exxon Crandon Zinc-

Copper Project," 13; <u>Ladysmith News</u>, August 6, 1981; <u>MJ</u>, September 18, 1980.

28.   Steve Brick, "Mining Policy in Wisconsin," April 26, 1982, 21-22, paper in the Mary Lou Munts Papers; Evans, "Development of Metallic Mining Policy in Wisconsin," 104; Wisconsin's Net Proceeds Tax on Mining and Distribution of Funds to Municipalities," April , 1997, WDNRP; James Derouin, speech, "The Wisconsin Model - A Consensus Approach," May 1, 1981, Peter Peshek Papers; League of Women Voters Questionnaire, #1, Munts Papers; OH, Linda Bochert.

29.   Ducey, "Welcome Senator," 16; Evans, "Development of Metallic Mining Policy in Wisconsin," 136; <u>Ladysmith News</u>, August 6, 1981.

30.   Douglas Amy, "Potential Political Problems in Wisconsin's 'Consensus Process,'" March 27, 1985, paper in the Mary Lou Munts Papers.

31.   Thomas J. Crawford to Roscoe and Evelyn Churchill, March 27, 1984, GRP.

32.   Douglas Amy, "Potential Political Problems in Wisconsin's 'Consensus Process,'" March 27, 1985, paper in the Mary Lou Munts Papers; OH, George Rock.

33.   <u>New York Times</u>, November 26, 1976; OH, Gordon P. Connor, OH, Sylvester Poler.

34.   Lester J. Burkett to Charles McGeshick; October 28, 1976, NMCP; <u>Proxy Issues Report</u>, April 28, 1983, GRP; <u>MCT</u>, July 16, 1977; OH, Arlyn Ackley.

35.   <u>MJ</u>, October 17, 1982; OH, Fred Ackley, OH, Sylvester Poler.

36.   Al Gedicks, "Activist Sociology," 57-58; OH, Arlyn Ackley, OH, Al Gedicks, OH, Daniel Poler, OH, Sylvester Poler.

37.   Green Bay Press-Gazette, June 8, 1980; OH, Daniel Poler, OH, Sylvester Poler.

38.   Al Gedicks, "Activist Sociology," 61-62; OH, Daniel Poler, OH, Sylvester Poler.

Chapter 2 Making Progress

1.   Clipping [1996], GRP; MJ, July 12, 1982; Russell, "Exxon Crandon Zinc-Copper Project," 8; OH, Barry Hansen, OH, Robert Russell.

2.   OH, Barry Hansen.

3.   Business Week, March 28, 1977; OH, Robert Russell.

4.   Green Bay News-Chronicle, February 18, 1985; MJ, October 17, 1982; Wisconsin Report, December 20, 1983.

5.   MJ, August 27, 1983; MS, November 9, 1983; WSJ, August 18, 1985.

6.   Crandon Report, April 1, 1983; Green Bay News-Chronicle, November 26, 1983; Northcountry Journal, April, 1985; WSJ, June 14, 1976.

7.   Mining in the North woods, January, 1985, GRP.

8.   Crandon Report, April, 1983.

9.   Extracts, September/October, 1985; Barry Hansen, speech, January 26, 1985, Barry Hansen Papers.

10.   Crandon Report, May, 1983; Barry Hansen, speech, April 6, 1984, Barry Hansen Papers.

11.   Crandon Report, June, 1983; Extracts, September/October, 1985; Barry Hansen, speech, January 26, 1985, Barry Hansen Papers; OH, Edie Franson.

12. *Business Week*, December 17, 1984; *Wall Street Journal*, January 4, 1978; *Proxy Issues Report*, April 28, 1983, GRP.

13. OH, Robert Russell, OH, J. Wiley Bragg.

14. Bill Tans to Ladies and Gentlemen, May 21, 1985, GRP; *Crandon Report*, April, 1984; *Mining Journal*, May 31, 1985; OH, J. Wiley Bragg.

15. Clipping, Wisconsin Legislative Reference Bureau Papers; Ducey, "Welcome Senator," 19-20; *MJ*, October 24, 1982 and November 24, 1983; Huffman, *Protectors of the Land*, 136, 162; Thompson, *History of Wisconsin*, 302; OH, Robert Ramharter.

16. Clipping, Wisconsin Legislative Reference Bureau Papers; Stan Druckenmiller to Kevin Lyons, January 12, 1979, Subject files, 1967-1987, Box 65, WDNRP/SHSW; Linda Reivitz to DNR staff, September, 22, 1980, Subject files, 1967-1987, Box 65, WDNRP/SHSW.

17. Robert Ramharter to Ladies and Gentlemen, March 14, 1986, GRP; DNR "Report," December 1982, GRP.

18. Larry Lynch to Gordon Reinke, April 25, 1984, Subject files, 1967-1987, Box 89, WDNRP/SHSW; Gordon Reinke to Barry Hansen, May 25, 1984, Subject files, 1967-1987, Box 88, WDNRP/SHSW; *Rhinelander Daily News*, April 13, 1984; *WSJ*, May 13, 1983.

19. OH, Stan Druckenmiller, OH, Robert Ramharter.

20. Stan Druckenmiller to Kip Cherry, February 23, 1978, Subject files, 1967-1987, Box 65, WDNRP/SHSW; *Wisconsin Report*, December 20, 1983; OH, Robert Ramharter.

21. OH, Stan Druckenmiller, OH, Robert Ramharter.

22.   Ducey, "Welcome Senator," 19-20; Appleton Post-Crescent, January 12, 1985; MJ, December 15, 1984.

Chapter 3 "All Tailings Ponds Leak"

1.   Clipping, GRP.

2.   Clipping, June 19, 1986, GRP; Green Bay News-Chronicle, February 18, 1985.

3.   Clipping, June 19, 1986, GRP; Green Bay News-Chronicle, November 26, 1983.

4.   Green Bay News-Chronicle, February 18, 1985; MS, December 7, 1982; OH, George Rock, OH, Jim Wise.

5.   George Rock to Tony Earl, October 7, 1983, GRP; MS, March 7, 1983.

6.   Mike Walter to George Rock, June 25, 1984, GRP; Evelyn Churchill to George Rock [no date], GRP.

7.   Clipping, GRP; Antigo Daily Journal, August 1, 1994; Ducey, "Welcome Senator," 20.

8.   Herb Buettner to Friends, April 11, 1985; SWP; Appleton Post Crescent, September 27, 1993 and October 26, 2003; Green Bay Press-Gazette, February 17, 1985.

9.   Wausau Daily Herald, June 24, 1997; OH, Sonny Wreczycki.

10.   Bill Tans to Sonny Wreczycki, December 5, 1994, WDNRP; OH, Sonny Wreczycki.

11.   Clipping, June 17, 1985, Wisconsin Legislative Reference Bureau Papers; Isthmus, June 6, 1986; MJ, October 24, 1982; OH, J. Wiley Bragg, OH, James Derouin, OH, Al Gedicks, OH, Barry Hansen, OH, George Meyer, OH, Peter Peshek.

12. Gedicks, "Activist Sociology," 59; OH, Al Gedicks.

13. Evelyn Churchill to Al Gedicks, February 7 [1984], GRP; Green Bay Press-Gazette, August 8, 1983; OH, Al Gedicks, OH, Sandy Lyon.

14. Kathleen Falk to Carroll Besadny, March 20, 1984, GRP.

15. Northcountry Journal, October, 1985; OH, Waltraud Arts.

16. Clipping, GRP; Orlando Sentinel, May 20, 1983; OH, Toni Harris.

17. Al Gedicks to George Rock, May 9, 1983, GRP; Richard Gutman to Cecilia Blye, November 30, 1984, GRP; clippings, GRP; Orlando Sentinel, May 20, 1983; OH, Arlyn Ackley, OH, Toni Harris.

18. Toni Harris to Arlyn Ackley, December 15, 1983, Toni Harris Papers; Al Gedicks to George Rock, March 30, 1984, GRP.

19. Toni Harris to Mary Ellen Butcher, February 9, 1984, GRP.

20. Toni Harris to Elliot Cattarulla, November 24, 1986, Toni Harris Papers; Proxy Statement, November 24, 1986, Toni Harris Papers; Green Bay Press-Gazette, May 12, 1985.

21. Al Gedicks, "Comments on Exxon's 'Forecast of Future Conditions,'" April 16, 1984, GRP; clipping, GRP; MJ, October 24, 1982; Wisconsin Report, December 20, 1983.

22. OH, Robert Ramharter, OH, George Rock, OH, Jim Wise.

23. Green Bay Press-Gazette, June 22, 1986.

24.   "Announcement of the Final Environmental Impact Statement Release," November 19, 1986, GRP; Final environmental impact statement, Exxon Coal...Wisconsin [1986], iv, 132, 201-207, WDNRP; Antigo Daily Journal, November 20, 1986.

25.   Final environmental impact statement, Exxon Coal...Wisconsin [1986], 201-206, WDNRP.

26.   Herb Buettner to Town of Nashville, October 1, 1986, GRP; Sonny Wreczycki to Stan Druckenmiller, May 20, 1986, SWP; Al Gedicks, "Comments on Exxon's 'Forecast of Future Conditions,'" April 16, 1984, GRP.

27.   Herb Buettner to Town of Nashville, October 1, 1986, GRP; Report, United States Department of the Interior, July 31, 1986, GRP.

28.   Green Bay Press-Gazette, December 11, 1986; MJ, December 11, 1986 and December 21, 1986; MS, December 11, 1986; OH, Edie Franson.

29.   MJ, December 11, 1986, and December 21, 1986; MS, December 11, 1986; Rhinelander Daily News, December 14, 1986.

30.   OH, Barry Hansen.

31.   Newsletter, Wisconsin Resources Protection Council, December 23, 1986; Green Bay Press-Gazette, December 11, 1986; MS, December 11, 1986.

32.   MS, December 11, 1986; WSJ, December 19, 1986.

33.   Green Bay Press-Gazette, December 15, 1986; OH, Gordon P. Connor.

34.   Clipping, Barry Hanson Papers; Ducey, "Welcome Senator," 16; OH, Robert Ramharter, OH, Confidential interview.

Chapter 4 Interim 1987-1993

1. <u>WSJ</u>, March 24, 1991; OH, George Rock.

2. Bill Tans to Annette Rasch, July 11, 1995, WDNRP; <u>Appleton Post-Crescent</u>, August 4, 1991; <u>Forest Republican</u>, August 18, 1994; <u>Green Bay News-Chronicle</u>, May 10, 2000.

3. Bill Tans to Roscoe Churchill, January 13, 1994, WDNRP; <u>Appleton Post-Crescent</u>, August 4, 1991.

4. <u>Green Bay Press-Gazette</u>, December 14, 1998; <u>Green Bay News-Chronicle</u>, May 10, 2000; <u>Pay Dirt</u>, October, 1998; <u>Wisconsin Engineer</u>, April, 1997.

5. Roscoe Churchill to George Rock [No date], GRP; <u>Eau Claire Leader-Telegram</u>, January 12, 1997; <u>Green Bay Press-Gazette</u>, December 14, 1998.

6. <u>Green Bay Press-Gazette</u>, December 14, 1998; <u>Green Bay News-Chronicle</u>, May 10, 2000; <u>WSJ</u>, March 20, 1994.

7. Mark Patronsky to Spencer Black, April 25, 1994, SWP; <u>MS</u>, July 11, 1988.

8. James Patton to Carroll Besadny, May 20, 1988, GRP; <u>MJ</u>, June 1, 1988; <u>MS</u>, July 11, 1988.

9. http://www.freeessays. cc/db/41/skx51.shtml.

10. <u>Chicago Tribune</u>, September 21, 1994; <u>Wall Street Journal</u>, June 26, 1989.

11. Schouweiler, <u>Exxon-Valdez</u>, 43, 55,57.

12. Clipping [1996], GRP; <u>Chicago Tribune</u>, September 21, 1994; <u>Wall Street Journal</u>, June 13 and 14, 1996.

13.　　Grinde, Native Americans, 164; Lurie, Wisconsin Indians, 70; Nesper, Walleye War, 3-4; Satz, Chippewa Treaty Rights, 93.

14.　　Treaty Rights: 2006 Edition, 29; OH, Fred Ackley.

15.　　http://www.alphacdc.com/treaty/bresette.html; Lurie, Wisconsin Indians, 70-71; Satz, Chippewa Treaty Rights, 93; Treaty Rights: 2006 Edition, 29-30; OH, Ron Seely.

16.　　"Economic Impact...in Wisconsin," 5, 10; MCT, January 12, 1997; Lake and Deller, "Socioeconomic Impacts," 1.

17.　　"Economic Impact...in Wisconsin," 5; Green Bay Press-Gazette, July 23, 1995; New York Times, December 26, 1994; Lurie, Wisconsin Indians, 80; OH, Gus Frank.

18.　　Grinde, Native Americans, 163; Grijalva, "Origins of EPA's Indian Program," 1-10, 97-99; Slade, "Environmental Regulation," 2; Wisman, "EPA History," 1.

19.　　Ulli Rath to Colin Macaulay, March 5, 1993, NMCP.

20.　　Ulli Rath to Colin Macaulay, March 5, 1993, NMCP; M.D. Krossey to H.T. John, February 17, 1987, NMCP; MS, September 15, 1993.

21.　　Ulli Rath to Colin Macaulay, March 5, 1993, NMCP.

Chapter 5 The New Crandon Mining Company

1.　　Clipping, GRP; MS, September 15, 1993, and May 14, 1995.

2.　　Appleton Post-Crescent, December 3, 1995; MCT, July 9/10, 1994; MS, November 22, 1993; WSJ, September 15, 1993.

3.　　Forest Republican, September 16, 1993; Pioneer Express, October 16, 1995; OH, J. Wiley Bragg.

4.   "Mining in Wisconsin," August, 1997; <u>Crandon Project Summary Update</u>, October, 1996, 9, 17; <u>WSJ</u>, October 7, 1995.

5.   "Mining in Wisconsin," August 1997; <u>Crandon Chronicle</u>, July 1996; <u>Racine Journal</u>, November 30, 1997.

6.   <u>Crandon Chronicle</u>, August and September, 1995, and January/February, 1997; <u>Crandon Project: Summary Update</u>, October, 1996; OH, Charles Curtis, OH, Lynn Smith.

7.   "Contributions Budget," December 31, 1998, NMCP; <u>Crandon Chronicle</u>, May, 1995; <u>MCT</u>, April 20, 1994.

8.   "Crandon Mining Company Community Case Study," July, 1995, NMCP; Mary Ann Pires to Jerry Goodrich, December 19, 1995, NMCP; <u>Crandon Chronicle</u>, November, 1995; <u>MCT</u>, April 20, 1994; <u>Nicolet News</u>, June, 1999; OH, Richard Diotte.

9.   Mary Ann Pires to Jerry Goodrich, September 5, 1995, and December 19, 1995, NMCP.

10.   "Crandon Mining Company Community Case Study," July, 1995, NMCP; Mary Ann Pires to Jerry Goodrich, December 19, 1995, NMCP.

11.   OH, Mike Monte.

12.   Janet Smith to Colonel James Scott, November 18, 1994, WDNRP; OH, Confidential interview.

13.   D.A. Cumming to M.S. Parett, February 7, 1996, NMCP; D.A. Cumming to File, February 15, 1996, NMCP; Jerry Goodrich to D.A. Cumming, February 20, 1996, and April 12, 1996, NMCP; OH, Jon Ahlness.

14.   Jerry Goodrich to D.A. Cumming, February 20, 1996, NMCP; "Cultural Resources Management," 1-2; OH, Jon Ahlness, OH, Sissel Johannessen.

15.   Charles Curtis to Rodney Harrill, August 6, 1996, NMCP; MJ-S, August 17, 1996; OH, Confidential interview.

16.   Charles Curtis to Rodney Harrill, August 6, 1996, NMCP.

17.   MJ-S, August 17 and 18, 1996.

18.   Peter Theo to Jerry Goodrich, October 20, 1995, NMCP.

19.   Jerry Goodrich to Bill Tans, August 26, 1994, NMCP; OH, Richard Diotte, OH, Confidential interview.

20.   J. Wiley Bragg to J.C. Landru, Jr., February 3, 1994, NMCP; OH, Tina Van Zile.

21.   Community Update, August 1996; OH, Rodney Harrill.

22.   MJ-S, May 25, 1997; WSJ, April 6, 2001; OH, Tina Van Zile.

23.   OH, Rodney Harrill.

24.   Clipping, GRP; MJ, February 19, 1995; WSJ February 16, and May 7, 1995.

25.   Clipping, GRP; Appleton Post-Crescent, December 3, 1995; WSJ, February 16, 1995; OH, Confidential interview.

26.   Clipping, GRP; OH, Kathleen Falk.

27.   Appleton Post-Crescent, December 3, 1995; OH, Larry Lynch, OH, George Meyer.

28. Appleton Post-Crescent, December 3, 1995; MJ-S, February 28, 1999; OH, Larry Lynch.

29. Don Moe to Chris Carlson, April 25, 1995, GRP; OH, Stan Druckenmiller, OH, George Meyer.

30. OH, Christopher Carlson.

31. WSJ, August 5, 1995.

32. DNR "Status Report," April, 1995, WDNRP; DNR "Status Report," October, 1996, WDNRP.

33. Bill Tans to Steve AveLallemant, February 26, 1997, WDNRP; The Press, May 4-10, 2000; OH, George Meyer.

34. OH, Stan Druckenmiller.

35. OH, Ken Fish.

36. Bill Tans to Len Pubanz, March 19, 1997, WDNRP; Transcript, Committee of the Whole Meeting, Natural Resources Board, August 20, 1996, WDNRP.

37. Roscoe Churchill to George Meyer, December 18, 1994, GRP; Dave Blouin to Larry Lynch, September 22, 1996, WDNRP; clipping, September/October, 1994, GRP.

38. Gale Wolf to Archie Wilson, March 2, 1998, WDNRP.

39. DNR Watch, October, 1998, WDNRP; OH, Larry Lynch, OH, George Meyer, OH, Gordon Reid.

40. WSJ, April 4, 2002; OH, George Meyer.

41. City Pages, April 19-25, 1996; OH, James Wood, OH, Confidential interview.

42.   George Meyer to Bart Olson, April 5, 1997, WDNRP; "Summary of Public Comments and Questions," July 21, 1997, GRP; OH, Jim Wise.

43.   Larry Lynch to George Meyer, July 18, 1996, GRP; Carl Zichella to George Meyer, April 28, 1997, WDNRP; <u>WSJ</u>, November 3, 1996.

44.   <u>City Pages</u>, April 19-25, 1996; OH, Jim Wise.

45.   <u>City Pages</u>, April 19-25, 1996; OH, Jim Wise.

46.   <u>WSJ</u>, November 3, 1996; OH, Jim Wise.

47.   <u>City Pages</u>, April 19-25, 1996; <u>Lakeland Times</u>, May 24, 1996.

48.   <u>City Pages</u>, April 19-25, 1996.

49.   <u>The Press</u>, May 4-10, 2000; <u>Shopper Stopper</u>, July 23, 1996; OH, Bart Olson, OH, Jim Wise.

50.   John Engler to Tommy Thompson, January 28, 1998, WDNRP; Tommy Thompson to John Engler, March 30, 1998, WDNRP; John Glenn to John H. Zirschky, March 12, 1998, WDNRP.

51.   <u>Antigo Journal</u>, August 13, 1997.

52.   Stan Druckenmiller to Rodney Harrill, April 14, 1997, WDNRP; Bill Tans to George Meyer, May 27, 1997, WDNRP; <u>WSJ</u>, March 26, 1996.

53.   OH, James Klauser.

54.   Mary Ann Pires to Jerry Goodrich, May 6, 1996, NMCP; Dale Alberts to Don Cumming, July 26, 1996, NMCP.

Chapter 6 "Save the Wolf River!"

1.   J. Wiley Bragg to Jerry Goodrich, November 27, 1993, NMCP; Appleton Post-Crescent, June 19, 1994; MS, September 16, 1993.

2.   Vernon Kincaid to Larry Lynch, May 19, 1994, GRP; WSJ, April 24, 1994; OH, Larry Lynch.

3.   Appleton Post-Crescent, December 3, 1995; New York Times, December 26, 1994.

4.   "Informational Meeting," Forest County Non-governmental Organization, April 20, 1994, SWP; news release, Midwest Treaty Network, February 3, 1994, GRP; Appleton Post-Crescent, April 24, 1994; Isthmus, June 27, 1997.

5.   John Teller to George Meyer, September 20, 1995, NMCP; "A Socio-Economic Impact Study of the Proposed Crandon Mine Project on the Menominee Indian Community," Menominee Treaty Rights and Mining Impacts Office, 1997, WDNRP.

6.   Transcript, Committee of the Whole Meeting, Natural Resources Board, August 20, 1996, WDNRP, OH, Ken Fish, OH, Sylvester Poler.

7.   Appleton Post-Crescent, June 27, 1995.

8.   OH, Don Cumming, OH, Ken Fish.

9.   OH, Gus Frank.

10.   Appleton Post-Crescent, June 19, 1994; Isthmus, June 24, 1994; MS, June 16, 1994; The Crandon Project: Summary Update, October 1996; clipping, GRP; OH, Linda Sturnot.

11.   OH, Tina Van Zile.

12.	OH, Arlyn Ackley, OH, Gordon Reid.

13.	Bob Deer to George Meyer and Craig Karr, September 25, 1997, WDNRP; OH, Arlyn Ackley, OH, Christopher Carlson.

14.	George Meyer to Valdas Adamkus, December 7, 1994, NMCP; Valdas Adamkus to Arlyn Ackley, September 29, 1995, NMCP; Gordon Reid to Jerry Sevick, July 25, 2002, NMCP; <u>Antigo Journal</u>, October 14, 1996; clipping, GRP; clipping, SWP.

15.	<u>Green Bay News-Chronicle</u>, May 10, 2000; clipping, GRP.

16.	Clippings, GRP; <u>MCT</u>, December 13, 1995; OH, George Rock.

17.	<u>WSJ</u>, January 14, 1998.

18.	Charles Norris to Albert Ettinger, January 12, 1998, GRP; press release, Midwest Office, Sierra Club, January 15, 1998, GRP.

19.	Douglas Cherkauer to Christopher Carlson, February 3, 1998, WDNRP; OH, Douglas Cherkauer.

20.	<u>Wausau Daily Herald</u>, October 17, 1997.

21.	<u>Appleton Post-Crescent</u>, April 24, 1994; <u>Isthmus</u>, April 17, 1998; <u>WSJ</u>, August 11, 1996.

22.	OH, Maureen Wreczycki, OH, Sonny Wreczycki.

23.	Dave Blouin to George Meyer, November 15, 1995, WDNRP; <u>Pioneer Express</u>, November 6, 1995; OH, Maureen Wreczycki, OH, Sonny Wreczycki.

24.	OH, Mike Monte, OH, Maureen Wreczycki, OH, Sonny Wreczycki.

25.   Clipping, GRP; OH, Toni Harris.

26.   Clipping, GRP; OH, Zolton Grossman, OH, Tina Van Zile.

27.   Clipping, GRP; OH, Zolton Grossman.

28.   Clipping, GRP; <u>City Pages</u>, April 30 - May 7, 1998; <u>Timber Producer</u>, July, 1998; OH, Confidential interview.

29.   OH, Dave Blouin.

30.   <u>Green Bay News-Chronicle</u>, May 10, 2000; <u>Shawano Evening Leader</u>, June 4, 1996; OH, Melinda Colucci, OH, Judy Pubanz, OH, Len Pubanz.

31.   Clipping, WDNRP; <u>Green Bay News-Chronicle</u>, May 10, 2000; <u>Isthmus</u>, June 27, 1997; <u>Timber Producer</u>, July 7, 1998; OH, Tom Soles.

32.   Grossman, "Let's Not Create," 153-154; OH, Ron Seely.

33.   <u>MCT</u>, May 10, 1997; <u>Wisconsin Resources Protection Council</u>, March 17, 1994.

34.   <u>Green Bay News-Chronicle</u>, May 10, 2000; <u>The Press</u>, May 4-10, 2000.

35.   <u>MJ-S</u>, February 23, 1997; OH, Mike Monte.

36.   Clipping, GRP; OH, Sandy Lyon, OH, Maureen Wreczycki.

37.   OH, George Rock, OH, Sonny Wreczycki.

38.   Zolton Grossman to George Rock [No date], GRP.

39.   Clipping, GRP; leaflet, "What You Can Do to Stop Exxon's Proposed Crandon Mine," May 5, 1997, WDNRP; OH, Carl Zichella.

40. Leaflet, "Resolutions Opposing the Crandon Mine or Pipeline," January 27, 1998, SWP; MJ-S, June 25, 1997.

41. Riemer, "Grass-Roots Power," 864-865; Isthmus, June 27, 1997; MJ-S, July 7, 1995; WSJ, July 8, 1997; OH, Richard Diotte.

42. Alice McCombs, EarthWINS, November 23, 2003; OH, Alice McCombs.

43. Riemer, "Grass-Roots Power," 860; Webster, "Barbarians," 40; WSJ, March 21, 1998.

44. Riemer, "Grass-Roots Power," 858-860.

45. OH, Rodney Harrill, OH, Gordon Reid.

46. George Reif to Bill Tans, October 23, 1997, WDNRP; Green Bay News-Chronicle, May 26, 1994, and October 5, 2002.

47. Gedicks, Resource Rebels, 174; Green Bay News-Chronicle, May 21, 1994.

48. Appleton Post-Crescent, April 10, 1997; OH, Confidential interview.

49. Jerry Goodrich to Don Cumming, April 12, 1996, NMCP.

50. MJ-S, November 18, 1996, December 8, 1996, and February 23, 1997.

51. Chicago Tribune, December 13, 1996; Pioneer Express, November 27, 1995, and November 25, 1996.

52. MJ-S, November 18, 1996; Pioneer Express, November 25, 1996; WSJ, January 30, 1997.

53.   Kevin Lyons to Richard Pitts, December 2, 1996, GRP;
        Chicago Tribune, December 13, 1996; Shepherd Express,
        December 12, 1996.

54.   Chicago Tribune, December 13, 1996; WSJ, December
        16, 1996.

55.   Al Gedicks to Mike Monte, January 8, 1997, GRP;
        Chicago Tribune, December 13, 1996; WSJ, December
        16, 1996.

56.   MCT, January 9, 1997; Rhinelander Daily News,
        November 16, 1997; OH, George Rock, OH, Sonny
        Wreczycki.

57.   Rhinelander Daily News, November 21, 1999; OH,
        Glenn Stoddard.

58.   News release, Survey Center, St. Norbert College, March
        1997, GRP.

59.   WSJ, August 11, 1996, March 8, 1997, and March 31,
        1998; OH, Christopher Carlson.

Chapter 7 Moratorium

1.   Wisconsin Resources Protection Council, Newsletter,
        June 1998, SWP; Isthmus, April 17, 1998; WSJ,
        December 9, 1994.

2.   DNR Watch, October 1998; City Pages, April 30 - May
        7, 1998; Timber Producer, July 7, 1998.

3.   Green Bay Press-Gazette, May 5, 1996; MJ-S, November
        16, 1997; Rhinelander Daily News, April 18, 1996; WSJ,
        April 30, 1996.

4.   MCT, October 20, 1997.

5.   Herb Reba to Bill Tans, September 14, 1996, WDNRP;
        WSJ, May 16, 1996.

6.   Dale Alberts to Don Cumming, July 26, 1996; <u>WSJ</u>, March 8, 1997.

7.   <u>Isthmus</u>, April 17, 1998; <u>MCT</u>, February 5, 1997; <u>WSJ</u>, February 3, 1998.

8.   Gedicks, <u>Resource Rebels</u>, 170.

9.   Rodney Harrill to Don Cumming, March 10, 1997, NMCP; news release, Nicolet Minerals Company, 1999, NMCP.

10.   Dale Alberts to Rodney Harrill, September 19, 1996, NMCP; Dale Alberts to Gordon Gray, March 13, 1997, NMCP.

11.   <u>MJ-S</u>, October 15, 1997; OH, Gerald Gunderson.

12.   <u>MJ-S</u>, February 18, 1997.

13.   Dale Alberts to Rodney Harrill, March 19, 1997, NMCP; Crandon Mining Company, Management Committee Meeting, Minutes, April 24, 1997, NMCP.

14.   Dale Alberts to Gordon Gray, March 13, 1997, NMCP.

15.   Rodney Harrill to File, July 24, 1997, NMCP.

16.   <u>MJ-S</u>, October 15, 1997; <u>WSJ</u>, October 15, 1997; OH, Gerald Gunderson.

17.   <u>WSJ</u>, March 8, 1997.

18.   <u>Appleton Post-Crescent</u>, January 22, 1998; <u>WSJ</u>, January 9, 1998, January 18, 1998, and March 31, 1998.

19.   <u>Appleton Post-Crescent</u>, January 22, 1998; <u>Isthmus</u>, April 17, 1998.

20.   <u>MCT</u>, February 16, 1998; <u>MJ-S</u>, February 5, 1998; <u>WSJ</u>, January 18, 1998.

21.   City Pages, April 30-May 7, 1998; MJ-S, April 23, 1998; Timber Producer, July 7, 1998; Wisconsin Resources Protection Council (Forest County), newsletter, June 1998.

22.   Bill Tans to Tom Ward, December 15, 1995, WDNRP; Archie Wilson to Bill Tans, October 29, 1998, WDNRP; DNR Watch, October, 1998.

23.   George Meyer to All DNR Employees, February 24, 1998, WDNRP; leaflet, "Moratorium Example Mines," February 1, 1999, GRP; MCT, February 11, 1999; Green Bay Press Gazette, July 16, 1998.

24.   "Misconceptions About Mining in Wisconsin," DNR newsletter, February 1998; DNR Report, November, 1998, WDNRP; DNR Report, April 1999, WDNRP.

25.   MJ-S, August 17, 1998; WSJ, February 7, 1998; OH, Confidential interview.

26.   OH, Robert Abel, OH, Confidential interview.

27.   Gordon Gray to Ulli Rath, March 26, 1997, NMCP; Corey Copeland to Gordon Gray, March 27, 1997, NMCP; Don Cumming to File, January 9, 1998, NMCP.

28.   Gordon Gray to Don Cumming, March 24, 1997, NMCP; Antigo Journal, January 24, 1998; MJ-S, February 2, 1998.

29.   Green Bay Press-Gazette, December 16, 1998; MJ-S, February 2, 1998; Nicolet News, April, 1998; clipping, GRP.

30.   Green Bay News-Chronicle, May 10, 2000; MJ-S, August 17, 1998; clipping, GRP.

31.   Mining Environmental Management, March 2000, 19-20.

32.   MJ-S, December 13, 1998; OH, Dale Alberts, OH, Don Cumming.

33.   Memo, "Nicolet Minerals Company and Mole Lake Sokaogon Chippewa Community," 2000, NMCP; Nicolet Minerals newsletter, February 2000; Mining Environmental Management, March, 2000, 19-20.

34.   Memo, "Nicolet Minerals Company and Mole Lake Sokaogon Chippewa Community, 2000, NMCP; OH, Don Cumming.

35.   OH, Confidential interview.

Chapter 8 New Owners, Same Frustration

1.   Chicago Tribune, September 18, 2002; Green Bay News-Chronicle, May 11, 2001; WSJ, October 1, 2001.

2.   Green Bay News-Chronicle, May 11, 2001; WSJ, October 1, 2001 and October 14, 2001.

3.   Green Bay News-Chronicle, May 11, 2001; MJ-S, October 14, 2001; WSJ, October 1, 2001; clipping, October 18, 2000, GRP.

4.   Green Bay News-Chronicle, November 7, 2001; MJ-S, June 1, 2001; WSJ, October 1, 2001; Wis.Politics.com, June 20, 2002; OH, Dale Alberts.

5.   WSJ, December 27, 2000, and October 1, 2001.

6.   Billiton news release, August 28, 2000, WDNRP; leaflet, Wisconsin Stewardship Network, January 27, 2001, WDNRP; WSJ, December 27, 2000.

7.   Zolton Grossman to Norton Tennille, November 3, 2000, SWP.

8.   Zolton Grossman to Sonny Wreczycki, October 30, 2000, SWP; news release, "Protesters Oppose Crandon

Mine...," May 19, 2001, WDNRP; Green Bay News-Chronicle, October 5, 2002; OH, George Rock.

9.   News release by Glenn Reynolds, September 2, 2002, GRP; WSJ, August 23, 2002 and September 3, 2002.

10.   Green Bay News-Chronicle, January 15, 2002, and June 21, 2002; Wis.Politics.com, June 20, 2002; OH, Stan Druckenmiller.

11.   Gordon Reid to Bill Tans, July 31, 1998, WDNRP.

12.   Donald Cumming to George Meyer, July 9, 1999, WDNRP; Green Bay News-Chronicle, May 10, 2000; OH, Dale Alberts, OH, Steve Kircher.

13.   Christopher Carlson to Gordon Reid, July 2, 1999, WDNRP; DNR Report, June 2001, WDNRP.

14.   MCT, May 7, 2001; Chicago Tribune, September 18, 2002.

15.   Dale Alberts to Darrell Bazell, December 16, 2002, WDNRP; OH, Don Cumming.

16.   Appleton-Post-Crescent, September 17, 2002; Green Bay News-Chronicle, September 18, 2002; MJ-S, October 1, 2002.

17.   Appleton Post-Crescent, September 17, 2002.

18.   Robert Imrie, Associated Press, September 18, 2002, WDNRP; Green Bay News-Chronicle, September 18, 2002.

Chapter 9 Sold to the Indians

1.   Forest Republican, April 16, 2003; MJ-S, April 12, 2003; OH, Gordon P. Connor.

2.   Green Bay News-Chronicle, April 19, 2003 and July 18, 2003; OH, Gordon P. Connor.

3.   Gordon P. Connor to Melissa DeVetter, May 29, 2003 and August 26, 2003, WDNRP; MCT, October 29, 2003; OH, Gordon P. Connor.

4.   Green Bay News-Chronicle, July 30, 2003; OH, Gordon P. Connor, OH, Glenn Reynolds.

5.   Earth First Journal, January-February, 2004; MJ-S, October 29, 2003; OH, Gordon P. Connor, OH, Glenn Reynolds, OH, Tina Van Zile.

6.   WSJ, October 28, 2003, and October 29, 2003; OH, George Meyer.

7.   Zolton Grossman to Sonny Wreczycki, November 3, 2003, SWP; Green Bay Press-Gazette, December 10, 2003, MJ-S; October 29, 2003; WSJ, October 28, 2003.

8.   Appleton Post-Crescent, October 29, 2003; OH, George Rock.

9.   WSJ, November 2, 2003; OH, Ron Seely.

BIBLIOGRAPHY

Manuscripts

J. Wiley Bragg Papers, Kingwood, Texas

Al Gedicks Papers, La Crosse, Wisconsin

Barry Hansen Papers, Littleton, Colorado

Toni Harris Papers, Sinsinawa, Wisconsin

Sandy Lyon Papers, Springbrook, Wisconsin

Mary Lou Munts Papers, Madison, Wisconsin

Nicolet Minerals Company Papers, Crandon, Wisconsin

Peter Peshek Papers, Madison, Wisconsin

George Rock Papers, White Lake, Wisconsin

Wisconsin Department of Justice Papers, State Historical Society of Wisconsin, Madison, Wisconsin

Wisconsin Department of Natural Resources Papers, State Historical Society of Wisconsin, Madison, Wisconsin

Wisconsin Department of Natural Resources Papers, Madison, Wisconsin

Wisconsin Legislative Reference Bureau Papers, Madison, Wisconsin

Sonny Wreczycki Papers, Pearson, Wisconsin

Carl Zichella Papers, Sacramento, California

Oral History Interviews

Abel, Robert

Achttien, Donald

Ackley, Arlyn

Ackley, Fred

Ahlness, Jon

Arts, Waltraud

Black, Ken

Blouin, Dave

Bochert, Linda

Bragg, J. Wiley

Buettner, Herbert

Carlson, Christopher

Carlson, Eugene

Cherkauer, Douglas

Christenson, Arlen

Churchill, Roscoe

Colucci, Melinda
(Naparalla)

Connor, Gordon P.

Cozza, Daniel

Cumming, Don

Curtis, Charles

Dawson, Tom

Derouin, James

Devetter, Melissa

Diotte, Richard

Druckenmiller, Stan

Falk, Kathleen

Fish, Ken

Frank, Harold (Gus)

Franson, Edie

Evans, Tom

Gedicks, Al

Grefe, Robert

Grossman, Zolton

Gunderson, Gerald

Hansen, Barry

Harrill, Rodney

Harris, Toni

Henschel, Kira	Poler, Daniel
Johannessen, Sissel	Poler, Sylvester
Kircher, Steve	Pubanz, Len
Klauser, James	Pubanz, Judy
Kunelius, Dave	Ramharter, Robert
Landru, Jim	Reid, Gordon
Lynch, Larry	Rock, George
Lyon, Sandy	Russell, Robert
Markart, Ken	Schmitz, Robert
May, Edwarde	Schroeder, Carlton
McCombs, Alice	Seely, Ron
McGeshick, Ray	Sevick, Jerry
Meyer, George	Soles, Thomas
Monte, Mike	Smith, Lynn
Mutter, John	Stoddard, Glenn
Olson, Bart	Sturnot, Linda
Ortloff, Jerry	Sutherland, Laura
Orwig, Lyle	Teller, John
Ostrom, Meredith	Terrell, Caryl
Peshek, Peter	Van Zile, Tina

Ward, Tom

Wise, Jim

Wood, James

Wreczycki, Edmund
(Sonny)

Wreczycki, Maureen

Zichella, Carl

One confidential
interview

Newspapers

Antigo Daily Journal

Appleton Post-Crescent

Business Week

City Pages (Wausau)

Daily Cardinal (University of Wisconsin-Madison)

Eau Claire Leader-Telegram

Environmental News

Forest Republican

Green Bay News-Chronicle

Green Bay Press-Gazette

Isthmus

La Crosse Tribune

Ladysmith News

Lakeland Times

Madison Capital Times

Masinaigan

Milwaukee Journal

Milwaukee Journal-Sentinel

Milwaukee Sentinel

Mining Journal

New York Times

The News (Shawano)

Orlando Sentinel

Pioneer Express (Crandon)

The Press (South Central Wisconsin)

Racine Journal

Rhinelander Daily News

Shawano Evening Leader

Sheboygan Press

Shepherd Express

Shopper Stopper

St. Paul Pioneer and Dispatch

Timber Producer

Wall Street Journal

Wausau Daily Herald

Wisconsin Outdoor News

Wisconsin State Journal

Newsletters

CBE Environmental Review

Center for Public Representation Bulletin

Crandon Chronicle

Crandon Project: Summary Update

Crandon Report

Community Update

DLAD Newsletter

DNR Watch

Earth First Journal

EarthWINS

Environmentally Concerned Citizens of Lakeland Area (ECCOLA)

Fighting Bob.com

Legislative/Executive Insight

Metallic Sulfide Mining Newsletter

Midwest Treaty Network

Mining Impact Coalition of Wisconsin

Mole Lake Environmental Newsletter

The Muir View

Nicolet News

North Country Anvil

Northcountry Journal

Wisconsin Bar Environmental Law News

Wisconsin Democrat

Wis.Politics.com

Wisconsin Resources Protection Council

Books

Erickson, Sue. Seasons of the Chippewa. Great Lakes Indian Fish and Wildlife Commission, 2000.

Gedicks, Al. Resource Rebels. Cambridge, MA: South End Press, 2001.

Gough, Robert. Farming the Cutover: A Social History of Northern Wisconsin, 1900-1940. University Press of Kansas, 1997.

Grinde, Donald and Johansen, Bruce. Ecocide of Native America. Sante Fe: Clear Light Publishers, 1995.

Grinde, Donald. (ed.) Native Americans. Washington, D.C.: CQ Press, 2002.

Hays, Samuel. A History of Environmental Politics Since 1945. University of Pittsburgh Press, 2000.

Huffman, Thomas. Protectors of the Land and Water. Chapel Hill: The University of North Carolina Press, 1994.

Kates, James. Planning a Wilderness: Regenerating the Great Lakes Cutover Region. Minneapolis: University of Minnesota Press, 2001.

Kline, Benjamin. First Along the River. San Francisco: Acada Books, 1997.

Lurie, Nancy. Wisconsin Indians. Madison: The Wisconsin Historical Society Press, 2002.

Merchant, Carolyn. The Columbia Guide to American Environmental History. New York: Columbia University Press, 2002.

Meyer, John M. (ed.). American Indians and U.S. Politics. Westport, Connecticut: Draeger, 2002.

Nesper, Larry. The Walleye War. Lincoln: University of Nebraska, 2002.

Nesper, Larry, Anna Willow, and Thomas King. The Mushgigagamongsebe District. 2002

Nichols, Roger. American Indians in U.S. History. Norman: University of Oklahoma Press, 2003.

Opie, John. Nature's Nation: An Environmental History of the United States. Fort Worth: Harcourt Brace, 1998.

Satz, Ronald. Chippewa Treaty Rights. Wisconsin Academy of Sciences, Arts and Letters, 1991.

Schouweiler, Tom. The Exxon-Valdez Oil Spill. San Diego: Lucent Books, 1991.

Smith, Duane. Mining America. Lawrence, Kansas: University Press of Kansas, 1987.

Stefoff, Rebecca. The American Environmental Movement. Facts on File, 1995.

Szasz, Andrew. EcoPopulism: Toxic Waste and the Movement for Environmental Justice. Minneapolis: University of Minnesota Press, 1994.

This is Exxon. New York: Exxon Corporation, 1981.

Thompson, William. The History of Wisconsin, Volume VI. Madison: State Historical Society of Wisconsin, 1988.

Treaty Rights: 2006 Edition. Odanah, Wisconsin: Great Lakes Indian Fish and Wildlife Commission, 2006.

Wilkinson, Charles. Blood Struggle. New York: W.W. Norton and Company, 2005.

Articles

Buege, Douglas. "The Crandon Mine Sage." Z Magazine Online 17 (February 2004) http://zmagsite, zmag.org/Feb2004/ buegepr0204. html.

Cook, James. "Exxon proves that big doesn't mean rigid." Forbes, April 29, 1985, 66-74.

"Cultural Resources Management." The U.S. Army Corps of Engineeers, St. Paul District. http://www.mvp.usace.army.mil/ history/default.asp?pageid-1175.

Ducey, Michael. "Welcome to the Global Economy, Senator."Corporate Report Wisconsin, February 1987, 14+.

"EPA Policy for the Administration of Environmental Programs on Indian Reservations." U.S. Environmental Protection Agency, American Indian Environmental Office. November 8, 1984. http://www.epa.gov/cgi-bin/epaprintonly.cgi

Gedicks, Al. "Activist Sociology: Personal Reflections." Sociological Imagination 33(1), (1996): 55-72

Gedicks, Al and Grossman, Zolton. "Tailing Exxon and Rio Algom." Multinational Monitor 16 (November 1995): 22-25.

Gedicks, Al and Grossman, Zolton. "Native Resistance to Multinational Mining Corporations in Wisconsin." Cultural Survival Quarterly 25 (Spring 2001): 9-11.

Grijalva, James. "The Origins of EPA's Indian Program" (March 28, 2006). SSRN Electronic Paper Collection. http://papers.ssrn.com/ sol3/papers.cfm?abstract-id=894103.

Grossman, Zolton. "Native and Environmental Grassroots Movements." Z Magazine 8 (1995); 42-50.

Grossman, Zolton. "Let's Not Create Evilness for This River." In Forging Radical Alliances Across Difference, edited by Jill M. Bystydzienski and Steven P. Schacht, 146-159. London: Rowman and Littlefield Publishers, 2001.

Kirkland, Richard. "Exxon Rededicates." Fortune, July 23, 1984, 28-32.

Lake, Amy and Deller, Steven. "The Socioeconomic Impacts of a Native American Casino." Madison: Agriculture and Applied Economics 403 (December 1996): 1-28.

May, Edwarde R., and Paul G. Schmidt. "The Discovery, Geology and Mineralogy of the Crandon Precambrian Massive Sulphide Deposit, Wisconsin." In <u>Precambrian Sulphide Deposits</u>, edited by R. W. Hutchinson, C.D. Spence and J.M. Franklin, 447-480. Waterloo, Ontario: International Standard Book, 1982.

Medlin-Hensch, Kira. "North woods Mining." <u>Conscious Choice</u> (September/October 1994): 16-17.

Peshek, Peter. "New Metal Mining in Wisconsin." <u>Wisconsin Academy Review</u> 24 (December, 1981): 24-27.

Reinke, Gordon H. "Wisconsin Environmental Regulation of Metal Mining." <u>Wisconsin Academy Review</u> 24 (December, 1981): 28-31.

Riemer, Jeffrey W. "Grass-Roots Power Through Internet Technology - The Case of the Crandon Mine." <u>Society and Natural Resources</u> 161 (2003): 853-868.

Rubin, Bruce. "The Crandon Mine: a Long, Long Road..." <u>Pay Dirt</u>, October 1998, 10-15.

Russell, Edmund III. "Lost among the parts per billion; Ecological Protection at the United States Environment Protection Agency, 1976-1993," <u>Environmental History</u>. 2(January, 1997), 29-51.

Russell, Robert. "Exxon Crandon Zinc-Copper Project, Wisconsin." <u>Skillings' Mining Review</u>, July 4, 1981, 7-14.

Slade, Lynn, etal. "Understanding Environmental Regulation on Indian Lands." November 3, 1999. http://www.modrall.com/articles/article_27_1.html.

Sproul, Mark W. "Boom Towns, Ghost Towns." <u>Mining Voice</u> September/October 1995): 39-43.

Threinen, C.W. "The Fabulous Wolf." <u>Wisconsin Conservation Bulletin</u>. 28 (March-April, 1963), 3-5.

Webster, Bob. "Barbarians at the Gates of Cyberspace." <u>Mining Voice</u> (January/February, 1998): 38-43.

Woessner, Robert. "Seven Generations." <u>Corporate Report Wisconsin</u>, May 1, 1996, 16-17.

Others

Evans, Thomas J., "An Analysis of the Development of Metallic Mining Policy in Wisconsin From 1965 to 1982," Ph.D. dissertation, University of Wisconsin-Madison, 1994.

"The Economic Impact of Native American Gaming in Wisconsin." Milwaukee: Wisconsin Policy Research Institute Report, 1995.

"Mining in Wisconsin," Supplement of <u>Corporate Report Wisconsin</u>, August 1997.